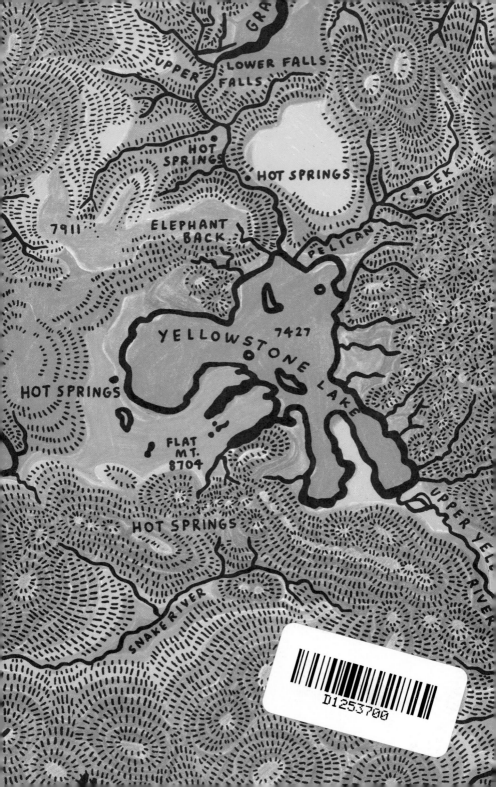

UPPER

LOWER FALLS
FALLS

GRA

HOT
SPRINGS

HOT SPRINGS

CREEK

7911

ELEPHANT
BACK

PELICAN

7427

YELLOWSTONE LAKE

HOT SPRINGS

FLAT
MT.
8704

UPPER YELL

HOT SPRINGS

RIVER

SNAKE RIVER

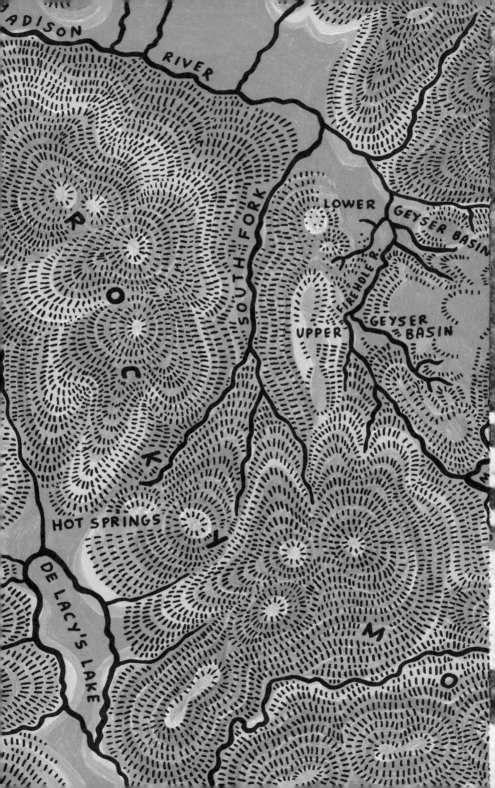

THE IMPERILED CUTTHROAT

Tracing the Fate of Yellowstone's Native Trout

GREG FRENCH

THE IMPERILED CUTTHROAT

Tracing the Fate of Yellowstone's Native Trout

GREG FRENCH

patagonia®

THE IMPERILED CUTTHROAT

Tracing the Fate of
Yellowstone's Native Trout

Patagonia publishes a select number of titles on wilderness, wildlife, and outdoor sports that inspire and restore a connection to the natural world.

Copyright 2016 Patagonia
Text © Greg French
Illustrations © Geoff Holstad

FIRST EDITION
Editors: John Dutton and John Paine
Art Director: Scott Massey
Illustrator: Geoff Holstad
Project Manager: Jennifer Patrick
Production: Rafael Dunn
and Jordan Damron

Printed in the United States of America
on 100% post-consumer waste

ISBN 978-1-938340-57-4
E-Book ISBN 978-1-938340-58-1
Library of Congress Control Number
2016906103

FOR THE
PLANET.
MEMBER

One percent of the sales from this
book go to the preservation and
restoration of the natural environment.

With gratitude to
Joe Brooks and Bob Behnke

Contents

INTRODUCTION 9

Chapter 1: The North Entrance 11

Chapter 2: Yellowstone Cutthroat Trout 37

Chapter 3: The Upper Lamar Valley 49

Chapter 4: Slough Creek 80

Chapter 5: Mammoth to Norris 92

Chapter 6: The Yellowstone River 102

Chapter 7: Yellowstone Lake 120

Chapter 8: The Snake River Catchment 142

Chapter 9: Back to Bozeman 169

Chapter 10: Angler Biologists 187

Chapter 11: Comparisons with Mongolia 200

Chapter 12: The Tasmanian Connection 218

Chapter 13: Lahontan Cutthroat Trout 233

Chapter 14: Endpieces 243

ACKNOWLEDGMENTS 255

INTRODUCTION

Beartooth Mountains, Yellowstone River, Absaroka Range, Sulphur Caldron, Buffalo Ford, Nez Perce, Firehole. The nomenclature of Yellowstone National Park would be a siren song even if you didn't know what it all stood for; I just wish I had happier reasons for surrendering to the call.

Greg French

Chapter 1:
The North Entrance

PARADISE

It is July 2012. Yesterday we flew from Tasmania to Montana and stayed overnight with Bob and Karin, whom we met while fishing for taimen on the Onon gol in Mongolia. This morning we have driven from Bozeman up Paradise Valley, alongside the Yellowstone River, into the Gallatin National Forest—through vast grasslands flecked with sparse copses of aspen and cottonwood, Douglas fir and Engelmann spruce. Frances is driving as we approach the famous "gate community" of Gardiner, the North Entrance to Yellowstone National Park. On the radio someone I don't know is singing Woody Guthrie's "This Land Is Your Land"—the popular version, the sanitized one.

I am looking at the Yellowstone map. I've been studying the contours and waterways for months. In my mind's eye I have wonderful three-dimensional images of every valley and ridge, basin and pinnacle; in my mind's nose, the scent of juniper and sagebrush, sulfur and buffalo shit.

The map emphasizes the extent of the greater Yellowstone wilderness. On the northern boundary of the national park I see where the Gallatin National Forest adjoins the Custer National Forest, incorporating the Sioux Ranger District. A bit farther away, on the eastern boundary, is the Shoshone National Forest. I love that name: shoh.**shoh**.nee. It reminds me of Nabokov's lo.**lee**.ta, though instead of tongue movements it's the whispers I find so seductive. There's nostalgia in the word too: as a child I was on the side of the Shoshone. Most of the time.

Greg French

I trace my finger over the southwest quadrant of the national park and settle on the Firehole River. There's a name that would captivate the most soulless traveler. Though perhaps you have to be a fly fisher to feel the butterflies I feel as my finger moves onward, downstream along the banks of the Madison. More interesting still is the northeast quadrant, especially the Lamar system.

Actually, everything ahead of us, and on all sides, seems inspiring and enchanting. We are heading somewhere mythical and, unfortunately, desperately sad.

"Where did you learn about Montana?" asked a visitor survey at the airport. "What made you decide to fly fish in Yellowstone National Park?" asked another researcher outside a tackle shop in Bozeman. The multiple-choice options on the Yellowstone National Park survey were no better: "What is the main activity you intend to undertake in the Park?" The list included hiking, nature study, photography, sightseeing, camping, and fishing. "Tick one box only."

People always want to know why you do what you do. Local newspapers, for example, often ask me to complete those brief quizzes routinely given to authors. I always find the question, *Which book influenced you the most?* mildly amusing. One book only?

Whenever I think about books, I am amazed at how many American titles I have come to adore. On the fishing side of things, there's pretty much everything by Joe Brooks, Robert Behnke, John Gierach, and David James Duncan. And on top of all that there's Twain's *Huckleberry Finn*, Steinbeck's *Cannery Row*, Hemingway's *The Old Man and the Sea*, Poe's *The Raven and Other Writings*—the list is endless. Even paperback westerns make the grade, though, like I say, I have always been on the Indians' side. Most of the time.

And how about comics? As a child I was enchanted by Carl Barks's epics. Do you remember Huey, Dewey, and Louie's *Junior Woodchucks Guidebook*? Today it seems like a hardbound precursor to the Internet, though it had infinitely more soul. The tale I remember most vividly was the one where Uncle Scrooge took his nephews *Back to the Klondike!* to look for gold. It was the first time anyone had ever suggested to me that bears might be anything other than cuddly and cute. I was amazed, too, by the chipmunks and moose. By the swarms of biting insects. By the mining history. By the extremes of weather. Everything was so unlike anything in Australia, and desperately compelling. But young as I was, I realized that Barks was really saying, "What is gold? Why is this yellow stone so valuable?"

Barks's comics were cautionary tales. In *The Golden Fleecing* and *The Golden Helmet*, which introduced me to the delights of mythology and history, he seemed to suggest knowledge itself could induce a type of gold fever; that knowing about the past did not necessarily empower you, that if you weren't careful it could compel you to perpetuate the mistakes of your predecessors.

Then again, maybe I have come to Yellowstone only because of Bob and Karin. "You will like Montana," they had insisted. "It's a lot like Mongolia." Well, Mongolia also reminds me of the Canterbury Plains in the South Island of New Zealand, and of the Patagonian steppes in South America.

If it's beginning to sound as if all of the world's great trouting destinations are alike, they are not. Tasmania, my home island, is quite possibly the most different place on Earth.

GARDINER

In *Travels with Charley*, a nonfiction account of his last road trip around the contiguous states, John Steinbeck declared that he held admiration, respect, and some affection for other American states but was in love only with Montana:

> "It seems to me that Montana is a great splash of grandeur. The scale is huge but not overpowering. The land is rich with grass and color, and the mountains are the kind I would create

13

if mountains were ever put on my agenda. Montana seems to me to be what a small boy would think Texas is like from hearing Texans."

Travels was published in 1962, the year of my birth, and during the drive from Livingston up through Paradise Valley I have been thinking of the difference between then and now. When I was born, Montana had a population of just over 700,000 people, which is not a whole lot different than the one million people it has today. To put it in perspective, Montana is five-and-a-half times as large as Tasmania but has only twice as many people, and Tasmania itself is mostly wilderness, production forest, and farmland. Yet Montanans are up in arms about the way Paradise Valley is being developed, the way rich out-of-staters have been buying up the ranches, building ugly mansions, making rural landscapes more suburban, restricting access. The situation is regrettable, no doubt about it, but for a visitor like me, the countryside still seems vast and sparsely populated.

I am especially surprised when we approach the outskirts of Gardiner, blink, and realize that we have managed to drive right through to the other side of town. Unexpectedly, we find ourselves at the foot of the Roosevelt Arch.

The Arch, bang-smack on the park boundary, is a rustic stone structure that has spanned the access road since 1903, when the railway was finally extended to Gardiner and the North Entrance became the first major gateway to Yellowstone. It's clunky and serves no practical purpose, yet I feel humble in its shadow. Perhaps this is just the ordinary power of an icon, even if I like to think I'm not often affected by such things. Maybe I like the way "FOR THE BENEFIT AND ENJOYMENT OF THE PEOPLE" has been plucked from the legislation that created the park in 1872 and literally written in stone. Of course, I am also happy that the arch carries Theodore Roosevelt's name.

Yesterday, in a coffee shop in Bozeman, we had a conversation with a fellow angler. He was impressed by our interest in Yellowstone and its history, and asked if we knew that Teddy Roosevelt had been a big-game hunter, or that Roosevelt's interest in conservation was piqued when he visited the Great Plains in the 1880s and saw firsthand the decimation of buffalo. Our newfound friend even pulled out his iPad and Googled the National Park Service website for us. He read aloud:

> "After he became President in 1901, Roosevelt used his authority to protect wildlife and public lands by creating the US Forest Service and establishing fifty-one Federal Bird Reservations, four National Game Preserves, 150 National Forests, five National Parks, and enabling the 1906 American Antiquities Act, which he used to proclaim eighteen National Monuments. During his presidency, Theodore Roosevelt protected approximately 230,000,000 acres of public land."

The same website also highlighted a famous Roosevelt quote, which was also read aloud:

> "It is also vandalism wantonly to destroy or to permit the destruction of what is beautiful in nature, whether it be a cliff, a forest, or a species of mammal or bird. Here in the United States we turn our rivers and streams into sewers and dumping-grounds, we pollute the air, we destroy forests, and exterminate fishes, birds and mammals—not to speak of vulgarizing charming landscapes with hideous advertisements. But at last it looks as if our people are awakening."

I said to the angler that I considered myself to be a natural constituent of the Democratic Party, and wasn't it funny that Roosevelt, a Republican, was my favorite American president?

"Well, the fact is," he replied, "most major conservation initiatives have been signed into law by Republicans. Did you know that it was Nixon who ticked off on the Endangered Species Act?"

"I guess we can be grateful that Roosevelt came first or else my favorite childhood toy might never have been called Teddy," I said.

He mulled words. "Dicky Bear. *Tricky* Dicky Bear. Nope, that wouldn't be much of a comfort to anyone."

Frances parks the car on the roadside near the arch, and we get out. Other cars stop so couples and families can stand at the foot of the monument to have their photo taken. We walk uphill away from the arch, across the open expanse of prairie, to gain a better view of the town, and I'm surprised to find small cacti, reminiscent of miniature prickly pears, interspersed with the grasses. From our slightly elevated vantage, Gardiner is dwarfed by the surrounding wilderness. The shop fronts look like a movie set for a spaghetti western. The town is so small that it's hard to credit 800 people live there.

We return to the foot of the arch, and Frances tells me to look up. Suddenly a voice at our side calls, "What's up there that you're looking at?"

"Swallows' nests," says Frances.

"Oh," says the man. "Well, you have a nice day, then."

The toll booth for the park lies just a few hundred yards past the arch. There is a steady flow of traffic, but no queue. Frances turns to me and says, "Breakfast in Gardiner?"

You bet. We need to go back anyway, for fishing licenses and advice.

Walking around the streets, looking at eateries and pubs, we quickly decide that we like this town. Despite its reliance on park tourism, the business center seems to be manned by locals and has a laid-back feel.

We stop outside a two-story edifice, rendered with cement and painted cream. It's a block construction, both materially and geometrically. Above the door, in large dark letters, are the words "Parks' Fly Shop." On the roof a neon sign says, "Park's Fly Shop." On the side wall a painted sign reads, "Parks Fly Shop." Clearly, the possessive apostrophe isn't sure whether it's working for Yellowstone National Park or Richard Parks, author of *Fishing Yellowstone National Park*.

We walk inside. A jovial man, short and rotund, greets us in the infectious Montanan manner, full of warmth and respect.

"Lovely town you have here," Frances says, and I back her up.

"New Zealanders," he deduces from our accent. He is so certain of the fact that we don't have the heart to correct him. At least he understands what we are saying; many Americans seem to have trouble with our thick Aussie vowels and the way we often make Ts sound like Ds, though they are always too polite to mention it. He asks us what he can do for us, and I say that we would like two park fishing licenses and some flies, locally tied if possible. What would he recommend?

"Well, they are all house patterns, and many are tied in-house." He points to a table near the door where a man sits creating exquisite grasshoppers.

The fly tier looks up and smiles. "Lovely day today, isn't it. Where are you going to fish?"

"Where do you think we should fish?"

"What sort of fishing are you after?" says the man behind the counter. "Dry fly? Nymphing? Wet fly?"

This strikes me as a quintessentially American line of questioning. I know that the stereotype is not properly representative; nonetheless, it is the way many American authors have chosen to set out their books. Even Joe Brooks did it.

It is so different from what I am used to at home in Tasmania and New Zealand. There, stalking trout is the name of the game, at least it is for me and my mates, and we rarely cast until we see the quarry. Sometimes the fish give away their positions by rising, sometimes by allowing their tails and dorsal fins to protrude from the water, sometimes by scattering schools of baitfish—but mostly we rely on polaroiding, which is local shorthand for using glare-defying polaroid sunglasses to spot fish swimming beneath the water's

Greg French

surface. "We tend to favor *functional flies* over perfect imitations," I say, paraphrasing Australia and New Zealand's most influential writer, Rob Sloane, author of the groundbreaking *The Truth About Trout* (1983).

"Functional flies?" says the man behind the counter.

"Well, for example, at home I can usually get nymphing fish to rise to a dry fly, providing the fly I use is either bulky enough to attract the fish's attention, or under tough conditions, sufficiently light so as not to spook the scales off it. The actual pattern doesn't seem to matter much to the fish, so a functional fly might be one that has a white post that *I* can see."

He nods but seems unconvinced, so I offer other examples. "In spring when brown trout are mooching about very shallow marshes feeding on frogs, they'll usually eat anything that makes a splash and wake. You don't need to waste time tying a fly that looks like a frog; you are better off putting your efforts into designing something that won't get snagged in the weeds. With crab feeders in our estuaries, the fly doesn't need to look much like a crab either, but it might have to be the right size and color."

"Well, that sort of attitude won't help you much here. You need to know where you're going to fish and what style you want to use so that you have exactly the right flies. You have to match the hatch to catch the fish."

"I guess that Down Under we don't always have good hatches, and maybe not the big concentrations of fish you have here, so maybe that's why we end up looking into the water so much and, when we spot a fish, making do with whatever fly we happen to have on at the time."

"Yeah, I've read about you Kiwis walking for miles up backcountry rivers in order to get just one or two chances in a day, and then probably not catching anything at all. Us? Well, if we fished for an hour or two and didn't get a fish we'd say, 'Well, we're bored now,' and go home. You got anywhere in mind that you want to go to?"

"Buffalo Ford?"

He shakes his head and sucks air through pursed lips. "Hardly any fish being caught there these days—because of the mackinaw problem in Yellowstone Lake, you know."

"Pelican Creek?"

He shakes and sucks like before. "I think it might be closed to angling. There are so few fish left in Yellowstone Lake that the spawning run that sustains Pelican Creek is in a real bad way."

"Heart Lake?"

"Long walk—a killer. And the problem with our lakes at this time of year is that the water's too hot and all the fish stay deep down in the middle. Don't waste your time. You're better off fishing for rainbows and browns in the Firehole and its tributaries, excepting that high temperatures might be a problem there too."

"I'm mainly interested in native fish. Anything happening at this end of the park? Lamar River?"

"The Lamar is just starting to clear. You won't do any good spotting, but prospecting the currents should be good. And if you're lucky you might get a hatch of pale morning duns, blue-winged olives, or even some caddis. Slough Creek will be worth a look too."

I always plan to immerse myself in the local culture and commonly vow to fish only according to local traditions, but I often disappoint myself. I went to Ireland in 2008 specifically to sample local lough-style techniques, and even then I could not stick to the task at hand. In recent years the weather had become unusually scatty: instead of the "normal" overcast skies, I had to contend with scudding clouds and lots of sun patches. Under these conditions the daphnia (planktonic water fleas) stayed deep, and the fish stayed deep too, way too deep to notice soft-hackled wet flies being skated about in the surface film. The locals persisted with tradition, but I soon stopped "prospecting" and began to concentrate on polaroiding individual fish, intercepting them with weighted nymphs.

We ask the man behind the counter to pick us a selection of flies, and then mosey around looking at books, maps, and décor. Why is there a bison head on the wall of a fly shop?

19

Afterward, we go to the Flying Pig camping store, and then we inspect the Sawtooth Deli. It turns out that we run into animal heads everywhere, even in the gift shop and visitor center. Everyone seems to be vying for the distinction of best dead head collection. Apparently you aren't in the running if you don't have at least one of everything big (bison, elk, pronghorn, bighorn sheep, Rocky Mountain goat, moose, mule deer), a random selection of little critters (marmot, rabbit, squirrel), and a jovial touch (a beaver with elk antlers, an otter with a snorkel, a hare with a deerstalker hat and popgun).

Standing in the shade of a Wild West–style verandah, which extends right out over the sidewalks, I decide that the chunky timber architecture is a genuine slice of Americana. I love it. Sometimes the external cladding is tacky—plastic logs and clapboard—but often enough it is real rough-sawn pine.

In the Blue Goose Saloon, sitting on wooden chairs at a wooden table, we glance around at the interior, typical of many eateries in Gardiner: troweled-on plaster with exposed timber beams. I ask if there is an Internet café in town, and the waitress says I can use her personal laptop. We order eggs, hash browns, and sausages. The serving is huge; otherwise it's fine. As for the coffee, it's weak, tepid, and overly bitter. There are many things they do well in America, but coffee isn't one of them. Maybe we should have said yes to sugar and creamer. I'd kill for some real loose-leaf tea.

Googling the current fishing conditions for Pelican Creek reminds me of how easy it has been to organize this trip to Yellowstone: everything from researching background information to booking accommodation and permits was done instantly from home with a few clicks of a mouse. When organizing my first major overseas trip—to Chile in the mid-1980s—the Internet didn't exist. The only guidebook I could access in Australia was the Lonely Planet, and for information on bushwalking I had to write to an independent American publisher and wait four months for a slim volume to return by post. As for fishing information, there was nothing available at all. These days my trips abroad are far more efficient, but I miss the serendipity.

By the time we finish Googling and eating, it is still only ten o'clock, and we decide we have time to buy some alcohol for the backcountry. The supermarket offers a great variety of bottled beer, including Trout Slayer, Salmon Fly Honey Rye, and Dancing Trout. The girl behind the counter is interested in our hiking plans and tells us about the "close encounter of the elk kind" her brother had last year. We pay for a six-pack of Moose Drool brown ale and a bottle of Flying Trout sauvignon (Cutthroat Blend). She hands back the change and says, "This your first time in America?"

"Sort of. We took our teenage kids to British Columbia in 2006 and had a transit in Hawai'i."

"How did you find customs and security in Hawai'i back then? To be honest, the way our border officials treat plane passengers, it's a wonder anyone ever comes back to the good old U.S. of A."

I tell her that we thought that the US security staff had watched too many cop shows.

"Well, you know, you were not alone. The Feds did a tourism survey and found that 90 percent of transit passengers felt the same way you did: Americans were unfriendly and America was no place to come back to. But 90 percent of visitors who had spent two weeks or more in the States felt that Americans were very friendly and everyone very much wanted to come back. The Feds took all sorts of steps to fix the image problem on the front line, and from what people tell me, it might be starting to work."

I admit that the staff at San Francisco International Airport a few days ago were exemplary, and I recall an edict on our US Customs Landing Card promising that customs and border protection officers would treat visitors in a "courteous, professional, and dignified manner," that "supervisors are available to answer your questions," and that "comment cards are available to compliment or provide feedback."

Greg French

CUTTHROAT

We have driven through the Roosevelt Arch, collected our park pass from the tollbooth, and are heading toward the town of Mammoth to pick up our backcountry permits for Pebble Creek and Cache Creek. I am supposed to be navigating for Frances, but I'm thinking about trout.

Did you know that each American state has a state fish? In Montana it's the cutthroat trout, elected by public poll in 1977. Over 200,000 Montanans thought it important enough to bother voting: not bad in a state that was at that time home to just 760,000 men, women, and children, a bigger percentage of voters than bothered to cast a ballot in the last presidential election. Everyone tells me that they preferred the cutthroat over other candidates because it was appropriately symbolic of clean, cold water; of healthy aquatic habitat; of Montana's natural heritage.

By way of contrast, Montana continues to market itself as "The Treasure State," as it has since gaining statehood in 1889. The official seal and state flag both depict the plow, pick, and shovel, and the state motto remains *Oro y Plata*, Spanish for "Gold and Silver." The legacy of mining is everywhere. As an angler you can't help but wince whenever you find out that an apparently beautiful stream is a designated Superfund site, a place so polluted by mine tailings that it has been officially earmarked by the US Environmental Protection Agency. Mind you, compelling anyone to do the cleanup always seems some way off.

I much prefer the state fish to the state motto. Heaps of other folk feel the same way, and not just Montanans. Very shortly, even before we get to Mammoth, we'll be in Wyoming, and in 1987, not to be outdone by Montana, the people of that state also chose the cutthroat trout as their state fish. Then in 1990, Idaho, the other state having territory within the park, adopted the cutthroat too.

I was eleven the first time I heard of the cutthroat trout. It was 1973 and I had borrowed a book—Joe Brooks's *Trout Fishing*—from my small-town library, where everything was on a two-week rotation from the city, never to be seen again once you returned it. I opened the cover and found page after page of glorious American salmonids that I never knew existed. I was intrigued by the Dolly Varden, steelhead, and grayling, but two fish came to dominate my imagination: the golden trout, which I hope to encounter soon enough, and the one with which I have recently become obsessed. The very name—cutthroat—must surely account for part of the attraction: it reeks of romance, fantasy, and danger.

Brooks even alluded to the very subspecies I seek today:

"Because the cutthroat trout is a wilderness trout rather than an urban dweller, he is one of the easier members of the family to catch, and his numbers have been decimated by heavy fishing wherever he can be reached. Today the Yellowstone River is probably the only stream outside of primitive areas which supports a good population."

Trout Fishing—like *Back to the Klondike!*—was utterly captivating. At the end of two weeks I could not bring myself to return it. First I pretended that I had mislaid it, then that my autistic brother must have "disappeared" it in the open fireplace or down the septic tank. Eventually, when I could stall no more, I admitted that I had lost it and paid the fine out of my own pocket money. My mother knew I was lying. Even the librarian eyeballed me over the top of her wire-framed glasses. Worse, I understood that paying for the book did not absolve me from having committed my crime. Forty years on, the guilt is still there. But I still won't give it up. If necessary, I would steal it again.

A raven caws, and I look up from the Yellowstone map. I see him on the roadside and he utters a maudlin "nevermore" before hopping thrice and flying off. I think of why I am here, and shudder.

Edgar Allan Poe is famous not just for "The Raven," but also for having introduced the Scandinavian word *maelström* into the

23

English language. And as if that wasn't enough, he practically invented the detective story. At least, *The Purloined Letter* and *The Murders in the Rue Morgue* were so perfect that their structure has come to define the genre. And now, one and a half centuries after these stories were written, Yellowstone National Park has offered up an American mystery as great as any Poe ever contrived.

Locally, the facts of the case are well known: in 1989, piscivorous lake trout, mackinaw to those of us who fish New Zealand, were illegally introduced into Yellowstone Lake and supposedly have eaten enough Yellowstone cutthroat trout to bring them to the brink of extinction.

The mysteries are manifold: Who is responsible for the mackinaw being there? Can anything be done? Will justice prevail? Common sense, even? Or is vengeance and face-saving more important than the outcome?

The supporting evidence poses many more questions than it purports to answer, and some months ago, I came to fancy myself as Poe's eccentric amateur sleuth on a certain path to outwit the official police force. That feeling, I'm nervous to admit, hasn't entirely gone away.

When I first heard that Yellowstone Lake had been illegally stocked, I assumed that the mackinaw had been liberated as fry or fingerlings. But chemical analysis of mackinaw ear-bones—on which the assumption of illegal stocking is based—is said to "prove" that the fish were liberated as adults in 1989. The never-discussed implication of this work is that an amateur angler somehow transferred hundreds, more likely thousands, of large mackinaw from neighboring Lewis Lake.

I know firsthand how hard it can be to collect large numbers of adult fish from the wild.

Tasmania's wild brown trout fisheries are so robust that stocks are usually limited by the carrying capacity of the lakes rather than recruitment: the more fish removed from the water, the greater the survival of naturally spawned fingerlings. For this reason the annual harvest of brown trout by recreational anglers in our most-fished waters has no measurable impact on fish stocks.

For half a century, Tasmanian fisheries managers have taken good advantage of this remarkable fact, transferring highland brown trout to lowland waters without adversely affecting the waters from which they have been sourced. And when liberated into our few understocked waters—mainly small dams without adequate spawning streams—these "wild transfers" have grown well and provided outstanding sport.

Traditionally, the only other widespread stocking was done by collecting ova from wild trout spawners and releasing the hatchlings before they started feeding. These fish were almost cost-free, essentially wild, and did well enough—especially when placed into waters that did not contain redfin perch.

In Tasmania in the late 1980s and early 1990s, I spent several winters working for the Inland Fisheries Commission collecting adult brown trout, and even though we had the luxury of official fish traps on major spawning creeks, we often struggled to collect the requisite number of trout.

By all accounts, adult mackinaw are more difficult to collect than adult brown trout. Typically, they live 60 to 120 feet below the surface and do not congregate in streams to spawn. How could an amateur fisherman catch so many of them? How could he keep them alive and transfer them nine miles by road from Lewis Lake to Yellowstone Lake? How did he avoid suspicion? (Not even the offer of a reward tempted anyone to come forward with information that may have led to "the capture and conviction of the person or persons responsible.") Why have none of the dozens of journalists who have reported on this problem for regional newspapers, fishing magazines, national broadsheets, and scientific journals alluded to this fundamental riddle?

Over the decades, while researching the various incarnations of my Tasmanian guidebook, I've needed to pinpoint unsanctioned releases of trout. The liberations themselves have invariably been simple operations: usually a cupful of sac-fry was purloined from a hatchery, or dip-netted from a creek, and placed into a plastic bag along with a few cups of water and a few quarts of oxygen; the fish were then stuffed into a backpack, walked into the wilderness, and liberated into a hitherto fish-free lake. Each of these tiny sorties was undertaken by one or two anglers on foot in untracked wilderness, and each, by its very nature, should have been easy to conceal, yet I usually got to the root of things within a few weeks or months. People like to boast, and I have found that if I talk to enough anglers, I get plenty of solid leads. What, I wonder, can be so hard about tracking down the perpetrator of an industrial-scale operation in one of the most-visited trout fisheries on the planet?

Perhaps because no one has been convicted, the villain of the story has become the mackinaw itself. As more than one observer has noted, "On the side of the lake trout is the lake trout. On the side of the cutthroat trout is everyone else."

The coalition of "everyone else" includes the US Geological Survey, National Park Service, US Fish and Wildlife Service, independent scientists, Trout Unlimited, independent angling clubs, the governments of Montana, Wyoming, and Idaho, and a plethora of conservation societies. So rarely do such disparate groups unanimously agree with one another, so rarely do even a few come together in a spirit of unity, that the Yellowstone disaster seems to offer hope—if not for the native fish, then at least for humanity—and this may be why the suppression of the mackinaw has become such a cause célèbre.

The tragedy came on the heels of a great success story. By the early 1970s, decades of overfishing in Yellowstone Lake and associated streams had resulted in catch rates that were by far the lowest on record. So angling regulations restricting harvest were introduced, then tweaked and tweaked again. Better still, the anglers themselves came to quickly embrace the catch-and-release ethos.

Fish numbers rebounded more quickly than anyone could have predicted. From the mid-1970s to the early 1990s, the lake was said to contain somewhere between 3.5 million and 4 million fish, which implied that the population had been restored to the pre-European norm.

But from 1994 (when the first mackinaw was caught) to 2004, the estimated population of lake-dwelling cutthroats dropped to perhaps 75,000 fish, which was much lower than the population in the early 1970s.

The published data suggests that Yellowstone Lake currently holds about 100,000 cutthroat trout, at best 400,000, and that they have to compete with perhaps 500,000 mackinaw. Even worse, I am told, each mackinaw is eating up to forty cutthroats per year, all of fingerling size or better.

I can't make the math work.

Another problem: the published data says that cutthroat trout from Yellowstone Lake spawn in rivers, and that when they drop back to the lake, the great majority feed in the shoreline shallows. This suggests that cutthroat trout and mackinaw mostly occupy completely different habitats. So when and where is the predation occurring? And why has it been so devastating?

Early on in my research, the thing that really struck me was that absolutely no one involved in the rehabilitation program—not the anglers, not the biologists, not the ecologists, not the conservationists—would countenance the resurrection of the local hatchery.

Now for another tantalizing fact: the year the mackinaw are supposed to have been illegally introduced into Yellowstone Lake, 1989, is the year after unprecedented wildfires ravaged the national park.

A number of pundits, including some who are not barking mad, have suggested that the mackinaw were inadvertently hoisted out of Lake Lewis by helicopters that were ferrying water buckets to fight the fires. Plenty of people have picked holes in this theory, however. For a start, so they say, the hotter the weather, the deeper the mackinaw, so what were the chances of picking up adults in buckets? And what were the chances of hundreds of fish flip-flopping across land that was at best parched, more likely smoldering, all the way into Yellowstone Lake's tributaries?

I was delighted to read that a few observers pointed an accusatory finger at waterbirds. I had a Carl Barks–like vision of an open-mouthed cartoon pelican flying high over the scorched Yellowstone wilderness, her beak full of innocent fish staring in goggle-eyed wonder at the smoky wasteland. Ah, well—every good detective story has to have its unlikely suspect.

I am starting to feel a bit grandiose again. This time I fancy myself not as an amateur sleuth but rather a type of Lone Ranger riding into town on a dark horse. I have vengeance on my mind. I lost my home water, my dear Lake Sorell, because of a carp-eradication project, and I'll be damned if I'll stand by and let this sort of thing happen elsewhere.

Not without a fight.

A much worse thought though, and a much more likely one, is that incompetence or a lack of imagination is *not* the problem. In Tasmania, I lost the Lagoon of Islands too, and the scientists working on that rehabilitation program were brilliant. Prolonged drought was found to be the culprit in that case, though when the drought persists through decades, you have to say it looks increasingly like climate change. What can anyone do about that?

I feel better when I have someone to blame and, really, like everyone else here in Yellowstone, I want to blame the mackinaw. If they truly are the problem, that will be the easiest thing to live with.

But, damn it, I don't want yet another childhood dream to wither! What scraps of my soul will be left to bequeath to my kids?

This is what I know from correspondence with local anglers and various people within Trout Unlimited and the National Park Service:

The spawning runs in Yellowstone Lake are in catastrophic decline, though some spawning streams are faring much better than others.

Catch rates are down (though not by the generally accepted 90 to 98 percent; it's actually more like 50 to 60 percent) and sizes are up.

Angling pressure has fallen dramatically at Yellowstone Lake and in the Yellowstone River downstream of the lake.

No one knows how the mackinaw were transferred into Yellowstone Lake, and it bothers me more than seems reasonable that the term "illegally stocked" is ubiquitous in published reporting, including scientific journals.

The suppression program has not suppressed the mackinaw. The population is said to be doubling every year and must soon climax, if it hasn't already.

For all of these reasons I am here one year earlier than planned.

Frances put up several legitimate arguments for sticking to our original schedule, whereby we were to incorporate this trip into an extended tour of the Americas in 2014. But the Yellowstone fishery was on a precipice, and if I had waited another couple of years, I might not have had the chance to experience it. Even if the Yellowstone cutthroat population remained high enough to sustain a viable fishery, what would happen if the population was listed as endangered under the Endangered Species Act? Would forty years of dreaming simply evaporate?

Right now I am filled with an urgency of the sort one feels when racing home to see an aging parent after a sudden illness. Or perhaps the urgency is more like the impatience I once felt for visiting undocumented wilderness fisheries in Tasmania—there is a great story to be told and I don't want anyone to beat me to it.

There is no doubt about it: the plight of the Yellowstone cutthroat trout has given me a passion for living that I've not experienced for years. I am grateful for it. And I find this gratitude for tragedy so disturbing that I sometimes wonder if any good at all can come from my being here.

MAMMOTH

The five major access roads, and figure-eight road system in the heart of the park, are all sealed two-laners. There isn't all that much traffic, not compared to what I had prepared myself for. Perhaps this is an illusion brought about because everyone goes at the mandated maximum pace (forty-five miles per hour) and there are lots of turnouts to view wildlife.

There are bugger-all Winnebagos (unlike the situation in British Columbia) so we always have a clear view of the road ahead. Instead, there is a superabundance of motorcycles. Most are Harley-Davidsons, and they usually travel in packs, sometimes as few as three or four, more often eight or twenty. Invariably the bikes are driven by middle-aged men, all heavily tattooed, most with pot bellies, some with ponytails. You can see their hair, or what is left of their hair, because no one wears a helmet. Or leathers. I am told that in America this is perfectly legal. Perhaps it is why biking is so much more popular here than in Australia.

Suddenly the traffic has slowed to a crawl.

Now we have completely stopped.

"A wildlife jam," Frances suggests.

I crane my head in front of Frances so that I can look out her side-window and up the hill that towers above us. Sure enough, a herd of twenty or more bighorn sheep has scattered itself about the steep slopes. Some animals are grazing, others stand sentry on rocky prominences. The cars ahead of us do their best to move to one side—leaving a single-car-width of room between the vehicles on the left and right—but no one drives ahead. Cameras appear from wound-down windows, but people mostly get out of their vehicles. And off their bikes.

"Awesome, isn't it," says a Kris Kristofferson look-alike to Frances as he and his female companion walk across the road for a closer look.

Whoop! Whoop!

A black and white patrol car has miraculously materialized. From a megaphone mounted on the roof, we can hear a ranger saying, "Now folks, it is important to stay seventy-five feet from all large wildlife." I feel like I'm on the set of *The Blues Brothers*.

Kris Kristofferson walks back toward his bike and gives the ranger a friendly wave.

"Thank you, sir," the ranger says politely into his megaphone, and by way of gratitude gives the siren another *Whoop! Whoop!*

I get out of the car for a better view, and a stranger standing on the side of the road smiles at me. Everyone is so damned friendly—I can't quite believe it. Nor can I believe that the roadsides are so clean. I watch the sheep. Look up into the blue, blue sky. Savor the warmth on my skin.

Eventually the vehicles begin to move on. A pickup overtakes our parked car from behind, and for a moment I'm convinced it is being driven by a golden Labrador. "The driver's seat is on the other side," says Kris Kristofferson, and then he gives me a jovial slap on the back. Apparently he is aware of my nationality.

A girl and boy, perhaps seven-year-old twins, walk past me and point excitedly to the number plate of the car in front of them. "Mom! Dad!" they yell in unison over their shoulders. "Kentucky! Only fifteen more states to find."

A huge biker remounts his bike and looks back toward Kris Kristofferson. "Let's go, Boo Boo." The sound of a fleet of revving Harleys reverberates harshly off the hillside, but it's hard to be offended when everyone is having so much fun.

The rest of the drive to Mammoth is remarkably straightforward. Again, I am surprised by how small and discreet the village is. Lonely Planet says the layout is "campus-like," and that's a pretty fair description. The buildings are made of stone, and the lawns are expansive and manicured. Most campuses don't feature so much wildlife, however; the public spaces are littered with elk, all cows, most lying about on the grass.

Mammoth was originally called Fort Yellowstone, being established in 1886, when the army had responsibility for park management. The name change didn't occur until after the Park Service was created in 1916. Why *Mammoth*? No one seems to know.

The building housing the Albright Visitor Center and Museum was built by the US Cavalry. It's an inconspicuous, charming two-story construction: the external walls built of sandstone blocks, the roof tiled in red terracotta. The main reception is refreshingly quaint, but more of a surprise is the backcountry office. Frances and I sit on plastic schoolroom chairs, waiting for the impossibly young ranger to finish a phone call. There is no one else in the room, which is lucky because it couldn't fit too many more. We stare at the tired paintwork, and I am reminded of my old dormitory at boarding school.

We booked our campsites three months ago and paid a reservation fee of twenty-five dollars per site. The actual permits are free, but they have to be collected in person no more than forty-eight hours ahead of the hike. And we have to do an induction, which is code for being forced to endure a "Summer Backcountry Video" titled *Beyond Road's End*. I've already seen it on YouTube. It was mildly entertaining, but only in a seditious way. It had relevant bits about bear safety, I guess, but like most other prospective hikers, Frances and I had read this sort of stuff a million times.

The bureaucracy involved in arranging an overnight hike in Yellowstone has been a real culture shock to me. At home in Tasmania we are free to walk in our national parks anywhere, anytime, and spontaneity has become central to the way we perceive the environment. Being a temperate-maritime island, our state experiences weather more volatile than that in Yellowstone, and intelligent decisions about where to fish can only be made at a moment's notice. If a high-pressure system suddenly pushes southeast off Central Australia, I'll probably decide the night before the trip that I want to go to a shallow clearwater where polaroiding will be

33

at its best. If a northeasterly system is going to bring heavy clouds, I'll probably opt for mayfly water. Heavy rain? Maybe a lake with suitable marshes and flood plains.

Also important is the ability to fish any destination we choose for however many days it takes to master difficult or unusual fishing situations, to not feel pressured before moving on to new challenges. We also like to fish the same locations in different seasons and under different weather conditions, mainly because we understand that this gives us an intimate understanding of the natural world that would be impossible if we were limited to visiting any one area for one or two days per year.

Bureaucrats have tried to rein us in, of course. From the late 1980s to the early 2000s, a small clique within the Tasmanian Parks and Wildlife Service tried to greatly restrict hiking and camping in Tasmania's World Heritage Area. They advocated a "carrying capacity" model of the sort which—since the 1960s—had been used to deal with the "negative impacts" of hikers in American reserves. The problem was, American researchers had already proven that, despite gobbling up the bulk of track funding, this sort of management program always failed to meet stated objectives. In fact, by the 1990s American land managers were strongly advocating alternative systems such as the Limits of Acceptable Change model, in which the stakeholders themselves decide how much change they are prepared to live with before restrictions are imposed. Unfortunately, users almost always want things to remain as they are "now," and such standards have a tendency to quickly become outdated, unsustainable, or irrelevant. The number of people walking Tasmania's famous Overland Track was a few hundred per year in the 1950s; now it is more than 12,500. This increase in use has been made environmentally sustainable only by hardening the tracks and improving the huts. Old-timers wouldn't much like the "crowded" conditions of today, but current users wouldn't much like the infrastructure being ripped up or being told that they had a poor chance of winning a lottery to walk the track, either.

In Tasmania, the very idea of a restrictive quota system was eventually abandoned, and the funding for researching and implementing the quota system was subsequently redirected into "Priority Erosion Control" programs.

The ranger finishes his call, smiles nervously, and apologizes for the delay. I tell him about my YouTube experience. He apologizes for having to make us re-watch *Beyond Road's End*.

When the film is finished, he apologizes for making us fill out several forms. He apologizes again when I have to go outside to copy down the registration number of our rental car.

In the car park, armed with the ranger's pencil and clipboard, I manage to scare half a dozen tourists: "You aren't going to fine me for wrongful parking, are you?" I recount this story to the young ranger when I go back inside, and mercifully he doesn't apologize.

Finally we leave the military-style buildings and drive past the famous geothermal area, Mammoth Hot Springs. I start thinking about the historical interactions that the US Army had with Yellowstone's cutthroat trout.

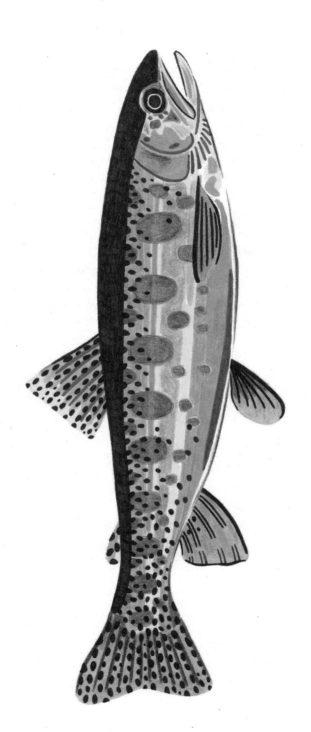

Westslope Cutthroat Trout
(*Oncorhynchus clarki lewisi*)

Chapter 2: Yellowstone Cutthroat Trout

SOLDIERING ON

What you probably remember about General George Armstrong Custer is that he and his entire battalion of more than 200 men were killed by the Sioux and Cheyenne at the Battle of the Little Bighorn on June 25, 1876. It seems that everyone in Montana and Wyoming remembers that stuff anyway.

What I remember about General George Armstrong Custer, other than the fact that his sepia photo-portrait in my childhood encyclopedia scared me witless, is that many of his troops were recreational anglers. Major Benteen, who commanded one of the two battalions that was not annihilated in the battle, was so keen an angler that Crazy Horse and Sitting Bull's braves knew him as *the officer on the big white horse who carries a fine fishing rod.*

As far as weathering war goes, General George Crook went one better than Major Benteen, discovering that you didn't need to survive the battle if you didn't actually turn up for it. A few days before Custer's defeat, Crook and his men were surprised by Crazy Horse and his braves on the south fork of Rosebud Creek, the first tributary of the Yellowstone downstream of the Bighorn, and forced to retreat, regroup, and fall back farther downstream to the next tributary, the Tongue.

So on the very day Crazy Horse dished out Custer's comeuppance, Crook's battalion went fishing. One man caught more than a hundred trout from Goose Creek, a tributary of the Tongue, and in the weeks that followed, Crook's troops harvested tens of thousands of trout from the Yellowstone and its feeder streams. *Tens of thousands*. Most fish were eaten, and apparently they provided a welcome respite from hardtack and pork.

There is not enough evidence to say if Yellowstone cutthroats were a significant food resource for local Indians in the same way that, farther south in Nevada, the Paiute relied upon the Lahontan cutthroat trout. Probably not. Within the Yellowstone National Park, at least, the winters were far too severe for permanent or semi-permanent encampments, even if numerous hunting parties traversed the region in summer. Nonetheless, the legendary American angler Edward Ringwood Hewitt, who fished in the park in 1882, mentioned giving trout he caught in the Snake River to a nearby band of Indians, and they seemed grateful enough.

YELLOWSTONE RIVER

We have just passed the administrative center of Tower-Roosevelt—even smaller than Mammoth—and are crossing the Yellowstone River. Below us is my quarry, though there is no time to fish here today.

The cutthroat as a species, as opposed to the Yellowstone race in particular, was the first American trout to be described by Europeans. In 1541 Francisco de Coronado, searching for one of the Indians' fabled cities of gold, found the Rio Grande cutthroat (*Oncorhynchus clarki virginalis*) in the upper Pecos River near present-day Santa Fe, New Mexico.

In 1805, captains Meriwether Lewis and William Clark, on their famous expedition of discovery—from St. Louis up the Missouri, across the Continental Divide, and down the Columbia to the Pacific coast—described the trout caught below the Great Falls on the Missouri River by one Silas Goodrich, a party member "remarkably fond of fishing." This subspecies of cutthroat is now known as the westslope, and its taxonomic name *Oncorhynchus clarki lewisi* honors the two expedition leaders.

The popular name—cutthroat—was first applied to what we now recognize as the Yellowstone subspecies. Charles Hallock, in an 1884 issue of *American Angler* magazine, wrote of fishing Rosebud Creek the previous year, noting that the trout he caught had "… a slash of intense carmine across each gill cover, as large as my little finger. It was most striking. For lack of a better description we call them 'Cut-throat' trout." Nowadays we know Hallock's fish as the Yellowstone cutthroat trout, *Oncorhynchus clarki bouvieri*.

Hallock's fish were in fact the exact same subspecies that General Crook's soldiers had caught in the Tongue and other Yellowstone tributaries. We can say this with confidence because the Yellowstone cutthroat is the only trout native to the Yellowstone system. If you want to know why, you'll have to endure a short geography lesson, but first a problem:

We writers of guidebooks invariably need to decide whether to describe a river from mouth to source or source to mouth, and I can tell you that neither way works particularly well. In the Antipodes, anglers almost always drive up a valley before getting out of the car and wading upstream, so in *Trout Waters of Tasmania* I opted for the mouth-to-source option. The advantage with this approach is that I often get to start at a big town on the estuary, usually one the reader will have heard of, perhaps one he intends to use as a base. The problem with this approach is that I am forced to describe lower reaches before the reader has developed an understanding of where the water comes from, or how it is affected by upstream tributaries, lakes, and dams.

Craig Mathews and Clayton Molinero, the authors of *The Yellowstone Fly-Fishing Guide,* which Bob and Karin sent us, opted for the source-to-mouth option. Their approach has the advantage of beginning at the beginning and ending at the end, and when they say something like, "The water that meanders between So-and-so Falls and So-and-so Rapids…," the reader can literally go with the flow. The disadvantage is that they have to first describe remote, itty-bitty headwater runnels of no particular interest to serious anglers

39

and hope that no one gets bored before they get to the juicy bits. And it doesn't resolve the problem of the reader having to perform mental gymnastics when traveling up the valley rather than down it. This is not a criticism of the book, and in Montana, Idaho, and Wyoming, where drift boating is a way of life, it makes more sense than it does at home. So for the moment I'll do it their way.

The Yellowstone River is located on the eastern or Atlantic side of the Rockies. From its source to its confluence with the Missouri, it flows nearly 700 miles, giving it the distinction of being the longest undammed river in the contiguous United States.

For reasons that will become obvious later on, I'll nominate as the source a minor tributary: remote North Two Ocean Creek in the Teton Wilderness, altitude 8,130 feet. Most texts, however, say that the source is technically farther downstream, just outside the southern boundary of the Yellowstone National Park, where the South Fork Yellowstone River merges with the North Fork Yellowstone River.

From the confluence of these forks, the Yellowstone proper winds north through the southeastern corner of the park across twelve-odd miles of wild plains flanked by marshes, bogs, and sloughs—a remote wilderness accessible only on foot. At altitude 7,732 feet it enters Yellowstone Lake, twenty miles long by fourteen wide. It exits the northern end of the lake at the Fishing Bridge and keeps flowing in a northerly direction for about fifteen miles, without losing much altitude, passing through the LeHardy's Rapids, Buffalo Ford, and Sulphur Caldron before roaring over the Upper Falls and Lower Falls into the Grand Canyon of the Yellowstone. From the Grand Canyon, the river passes through Black Canyon, picking up more water from the Lamar River, and exits the northern end of the national park at Gardiner.

The Yellowstone emerges from the foothills at Livingston, a mere 4,501 feet above sea level, and swings to the east, where it crosses rolling farmland and is fed by a succession of small but well-known trout streams, including the Stillwater and Boulder. Still farther down it is joined by Clarks Fork, and past Billings it picks up the Bighorn, Rosebud, and Tongue, all of which are sourced from mountains to the south.

Just below the confluence of the Tongue, at Miles City, we come to the point where Captain John Bourke, one of General George Crook's aides and diarists, attested the natural downstream limit of the Yellowstone cutthroat trout.

Beyond Miles City the Yellowstone meanders across hotter, drier prairieland and is fed by fewer coldwater tributaries. The water slows and warms and becomes more suited to catfish, paddlefish, burbot, smallmouth bass, and walleye. It eventually joins the Missouri, which flows southeast into the Mississippi, which in turn flows south into the Gulf of Mexico, by which time the water that fell in North Two Ocean Creek has pretty much passed across the full latitude of the contiguous states.

EVOLUTION OF THE YELLOWSTONE CUTTHROAT

The genus *Oncorhynchus* evolved on, and dominates, the western side of the Rockies in the Pacific drainage. How then did *Oncorhynchus clarki bouvieri* get up over the Continental Divide and into the Atlantic drainage?

I began researching this problem many months before my arrival in America, and one scientist's name cropped up again and again, often in the most unexpected places. Professor Robert J. Behnke had authored or coauthored many of the papers I was reading. In other papers he was credited by the authors for supplying technical advice or for properly contextualizing their findings. Many species and subspecies of trout had been named by him. At least one had been named after him. But what was most remarkable, I came to understand, was the man's uncanny ability to identify the crux of

41

any ecological or taxonomical problem and to come up with theoretical answers that almost always ended up being right regardless of how left-field they seemed at the time.

I recognized a problem: any in-depth story about Yellowstone's trout was in danger of becoming a biography on Behnke. Rather than fight it, I arranged an interview with the man, and will travel to Fort Collins in Colorado at the end of our Yellowstone trip to meet him. In the meantime I've made a point of reading and rereading his body of work, and among the books I am carrying with me on this trip is *Trout and Salmon of North America*, which provides a perfect refresher course on the evolution of America's most prized fishes.

About 30 million to 25 million years ago, a common ancestor of the subfamily Salmoninae gave rise to a char and taimen lineage and a trout and salmon lineage.

Perhaps 20 million to 15 million years ago, the trout and salmon lineage split into two different trout and salmon lineages: one population confining itself to the North Atlantic Ocean and becoming genus *Salmo*, the other confining itself to the North Pacific and becoming genus *Oncorhynchus*.

Five million years ago, the North Pacific population of the trout and salmon lineage began to divide into Pacific trout and Pacific salmon.

Two million years ago, the Pacific trout divided into what have become the rainbow trout (generally with 58 to 60 chromosomes) and coastal cutthroat trout (always 68 chromosomes).

A little more than a million years ago, the coastal cutthroat trout diverged, giving rise to a new subspecies: the ancestral westslope cutthroat (66 chromosomes).

A short time later, the westslope cutthroat diverged, giving rise to two more lineages: the Yellowstone cutthroat (64 chromosomes) and the Lahontan cutthroat (also 64 chromosomes).

Thus we have four major (ancient) lineages of cutthroat, though over the last 100,000 years some populations of the Yellowstone and

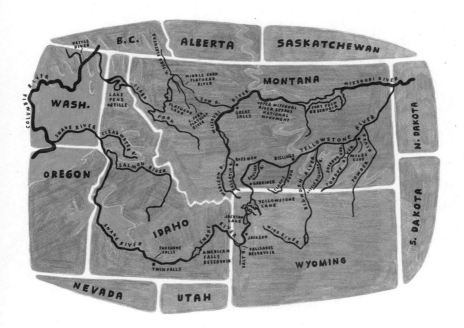

Lahontan varieties subsequently became isolated above barrier falls or found their way into adjacent catchments. Behnke recognizes one major and six derivative subspecies in the Yellowstone cutthroat lineage, and one major and four derivative subspecies in the Lahontan cutthroat lineage. (There is only one type of coastal cutthroat and only one type of westslope cutthroat.)

Behnke surmises that the Yellowstone cutthroat evolved from a population of ancestral westslope trout that became isolated in the Snake River catchment. Some 60,000 years ago—well after the last major eruptions and lava flows occurred farther upstream in what is now Yellowstone National Park—the Shoshone Falls came into being, dividing the river into separate lower and upper sections. Sometime after that, rainbow trout invaded the Snake system up as far as the falls, completely replacing the Yellowstone cutthroat in the lower section. Yellowstone cutthroat persisted above Shoshone Falls only because they remained isolated from all other trout species.

Then, perhaps just 6,000 years ago, after the last of the glacial ice retreated from the headwaters of the Snake River drainage, the Yellowstone Plateau was exposed and a remarkable geological curiosity occurred. At Two Oceans Pass, deep within the wilderness of the Bridger-Teton National Forest (part of the Greater Yellowstone Ecosystem) tiny North Two Ocean Creek split perfectly in two, forming Pacific Creek (which flows into the Snake River and the Pacific Ocean) and Atlantic Creek (which flows into the Yellowstone River and the Atlantic Ocean). Yellowstone cutthroats immediately swam up Pacific Creek and down Atlantic Creek, and soon colonized the entire Yellowstone catchment down to Miles City.

Even today, decent-sized fish should have no trouble swimming from Pacific Creek into Atlantic Creek and vice versa. I say *should* because Behnke points out that mostly they don't. In North Two Ocean Creek and Pacific Creek, the Yellowstone cutthroats are nonmigratory, and fish of many different ages are always present. In Atlantic Creek, adult fish are found only when the spawning run from Yellowstone Lake is under way, and thereafter you pretty much

45

only find fry and fingerlings. It seems that, these days, differences in behavior and life history largely keep the Pacific Creek and Atlantic Creek populations isolated from one another.

I don't know of any other place in the world where it is currently possible, theoretically or otherwise, for fish to cross back and forth across a continental divide, though there are countless places where, in the geological past, upheavals have tipped one catchment into another, and where rivers have been dammed by ice or landslides and been diverted over ridges. Indeed, the westslope cutthroats that Silas Goodrich caught below the Great Falls on the Missouri River are the result of an ancient geological transfer from the Flathead system (in the headwaters of the Pacific-flowing Columbia River basin).

UPHEAVAL

Yellowstone Lake itself has a history of upheaval and subsidence. Indeed it is thought that in the dim past, so long ago that no trout at all existed in the upper Snake and Yellowstone systems, the original Yellowstone Lake actually drained into the Snake River.

Even today the lake remains extraordinarily volatile. In 1973—the year I stole *Trout Fishing*—Professor Bob Smith, a geologist, was camped on Peale Island at the southern end of the South Arm of Yellowstone Lake when he suddenly realized that the dock (more of a jetty if you are Australian) was underwater and that forests and a meadow flanking the southern shores had been inundated by a foot or more of water. Later, when he returned to the north of the lake, he noticed that the northwestern shore had receded, but the outlet appeared unchanged. The only plausible explanation was that the land beneath the lake had uplifted and tilted. Smith also knew that major eruptions of the Yellowstone volcano had been occurring every 600,000 years, and that the last one had occurred 640,000 years ago. It was very unsettling.

Just a few years before Smith's observations, another famous geologist, Bob Christiansen, had become troubled by the fact that no one had yet found the whereabouts of the Yellowstone volcano. He knew it had to be somewhere: sulfur cauldrons and fireholes and Old Faithfuls don't occur for no reason. Then a considerate NASA employee donated a high-altitude photograph of the Yellowstone region to the National Park Authority for possible display in a visitor center, and Christiansen saw the reason the volcano had never been found. Simply, it was too big to be noticed from the ground. Or even from a helicopter. To be exact, it was forty-five miles long by thirty-four wide. This was not just any old volcano: it was a supervolcano, the biggest on the planet. Furthermore, it was located bang-smack in the middle of the park, though this did not really surprise Christiansen since the park boundaries had been drawn specifically to preserve every single geothermal feature.

The type of crater Christiansen was looking at is known as a *caldera*, Spanish for cauldron. We anglers of the Antipodes know all about them because Lake Taupo on the North Island of New Zealand—the largest lake in the country, the nation's greatest trout fishery—is a water-filled caldera.

Calderas are formed in a way quite different from that of ordinary volcanic craters. A fissure in the Earth's crust allows some magma to rise toward the surface, but then the magma gets trapped beneath a hard skin of rock. Pressure builds, and an angry blister forms. If the pressure continues to build, the perimeter of the blister ruptures and you get a major volcanic eruption. But the skin itself is usually too thick and heavy to be ejected, and once the magma chamber is emptied, it sinks down into the resulting cavern. But that's not necessarily the end of the matter. Often the magma chamber starts refilling, as has been the case time and time again at Yellowstone (and Taupo); as is the case today.

Supervolcanos do quite a bit of damage when they explode. When the last major event of this type occurred, at Toba on the Indonesian island of Sumatra some 74,000 years ago, the ash cloud

47

resulted in six years of global volcanic winter. *Homo sapiens* was taken to the very brink of extinction; it is thought that all but a few thousand people survived on planet Earth.

The Yellowstone volcano is much, much bigger than the one at Toba, and the last time it erupted, ash rained down upon all of the United States west of the Mississippi, and large parts of Canada and Mexico too. In many places, it piled up six feet deep. Assuming that anyone survives the next blast, they can forget about waiting for the dire effects of a massive volcanic winter: with most of its crops destroyed, America would quickly starve. Actually, the whole planet would starve because the western part of the United States produces fully half of all the world's cereal.

The current blister beneath Yellowstone has been festering for 640,000 years. It's not just red and angry, it's throbbing. Literally. The caldera rose from 1973 until the mid-1980s, subsided until the mid-1990s, and then began rising again. Recently it has been heaving up and down at whim.

Thousands of tiny earthquakes occur every year. Most are too small to be felt, but there was a fatal shaker of magnitude 7.5 at Hebgen Lake (just outside the west boundary of the national park in Montana at the head of the Madison River) in 1959 and a non-fatal one of magnitude 6.1 inside the caldera in 1975.

Chapter 3: The Upper Lamar Valley

LAMAR VALLEY

The Lamar Valley, we have been told, is Yellowstone's Serengeti, and sure enough, as soon as the plains come into view, we find ourselves staring in goggled-eyed wonder at vast herds of bison, or buffalo.

I have trouble with the word *buffalo*. Perhaps it's because in Australia it applies to an altogether different beast, the feral Asian water buffalo. Perhaps it's because *buffalo* seems too bland a word to use for a creature so prehistoric, so mystical. Perhaps it's because *buffalo* undermines the bison's ecological importance. I concede, however, that the word *buffalo* has undeniable historical and cultural legitimacy here in America.

We turn into a pullout where several other cars are parked. I open my door and the smell of buffalo shit and sagebrush hangs comfortingly in the air, as familiar to my imagination as to a ranger's nose. A man has set up a spotting scope, and he invites us over. "Do you want to see a wolf? You'll have to be quick now."

Frances races over and glimpses the wolf just as it disappears over a distant ridge. "Cool!" she says.

"You're welcome," says the wildlife aficionado, grateful for her enthusiasm.

Back on the plains, the closest group of about a hundred buffalo has decided to migrate to the other side of the highway, and the leading bulls are approaching the road verge. I'm told that a large male buffalo can weigh almost a ton, run more than twenty miles per hour, jump more than six feet vertically from a standing start, and swim a bit. We begin to retreat back to the car. Other people raise their cameras and move in for a closer look.

49

Whoop! Whoop! "Now, folks, move away seventy-five feet from all large wildlife. You, sir. Yes, you too. Thank you." *Whoop! Whoop!*

How do these ranger vehicles manage to arrive on the scene so quickly?

By now buffalo are milling around our car. Bikers seated on stationary Harleys switch off their engines and smile to one another. They keep smiling even when some of the animals brush against their exposed arms. Tentatively I wind down the window. Some animals are grunting and snorting. Others are making a soft drumbeat deep in their throats, much like emus. The biggest bull is bloody enormous, much bigger than any full-grown Angus or Hereford I've seen. Or even a Murray Grey.

Finally we are able to drive ahead, and soon the road draws close to the true right bank of the Lamar. I convince Frances that we have time for a quick look at the water, and we pull in to a turnout. There are a couple of fly fishermen down there, but plenty of room for newcomers. If we care to walk just a few hundred yards upstream, we will be completely alone. Alone except for another herd of buffalo.

As we kit up and walk down to the water, I think about Zane Grey's *The Thundering Herd*, the 1925 western novel in which he expressed melancholy over the demise of buffalo. The movie based on the book, a Randolph Scott vehicle released in 1933, was filmed right here in the Lamar Valley, with the park rangers staging the stampede. I have a soft spot for Grey's westerns, not because I think he wrote them particularly well, but because in Australasia he looms so large in our fishing lore. His expeditions to New Zealand in the mid-1920s and early 1930s were extraordinarily influential in the United States, mainly because he wrote so passionately about them in books and major outdoor magazines. He was primarily interested in game fish (he practically invented the teaser and wrote extensively of the Bay of Islands), but he also loved the Tongariro—the major tributary of Taupo, the caldera lake—describing it as the greatest trout stream in the world. I was especially impressed with his determination to go beyond the nuts and bolts of fishing to comment on culture, landscape, and wildlife. I was completely pissed off

with the unflattering biography written by amateur Kiwi histori-
an Bryn Hammond in *Fish&Game* magazine. Hammond's main
problem seemed to be that he didn't like foreigners catching "his"
fish. Another was that Grey's lifestyle offended his twee morality.

Suddenly we are at the water's edge and my frustration abates.

The Lamar is a mid-sized freestone river, not too much smaller
than the Tongariro, which is to say that most pools are less that
forty yards wide and you can wade across many of the tailouts.
The water is discolored—a gray turbidity that limits polaroiding to
a depth of three feet at most. There are no fish on the edges, and no
rises. On the other hand, I spot obvious current seams where fish
should be stationed. We tie on dry attractors—grasshoppers—and
begin fishing.

After an hour of prospecting, we still haven't had a hit. I thought
cutthroats were supposed to be pushovers. Reluctantly I tie on a
heavy nymph dropper and instantly get a few flighty takes from
what I assume to be sprats.

Ten minutes later, I land an eight-inch trout, my first-ever
Yellowstone cutthroat. Then I hear Frances, fifty yards behind me
on the opposite side of the river, telling me to look up. An enormous
herd of buffalo has coalesced around me.

Strangely enough, considering how nervous I am around domestic bulls, the buffalo don't really bother me. Maybe that's just
because I feel that I am on the verge of a fishing bonanza.

"Look out for the cripple!" Frances warns, and I see a cranky
bull buffalo with a withered leg wading across the river toward me.

"He'll be wolf tucker come winter!" I yell back, but I back off
downstream anyway, and Frances and I begin dodging our way
through the milling herd, rarely if ever able to keep the mandatory
seventy-five feet from the nearest animals.

In Tasmania, walking home through a patch of grazing kangaroos or wallabies doesn't raise the heartbeat in quite the same way.
Actually, nothing in the Tasmanian bush is going to hurt you: not
the pademelons, not the bettongs, potoroos, wombats, or platypuses.
Certainly not the sugar gliders or pygmy possums.

Greg French

The Tasmanian devil? Sure, it's carnivorous, but despite what Warner Brothers may have suggested, it's welcome around any campsite. The worst that might happen is that your backpack gets raided while you sleep, or a cheeky youngster runs away with a pot or pan.

Snakes? Admittedly there are an awful lot of them, and they're all poisonous, but they're timid and hardly likely to attack unless severely provoked.

I start to think about other differences. In Tasmania we don't have prairies; mostly we have trees—primal rainforests, wet sclerophyll forests, eucalypt woodlands. There are plains in the Central Highlands, but they are smallish and moor-like.

The moors are, in fact, my spiritual home, especially the ones in the Western Lakes wilderness. And this makes me think of another big difference between here and home. All the trout lakes and rivers in Yellowstone have long been written about, and each has its own lore. When I was a teenager, hardly any of Tasmania's backcountry fisheries had been written about; some may never have been fished. My homeland is so wild that I had the opportunity to be the first person to document many of the waters I explored. I wonder if such places still exist anywhere in Yellowstone. Or anywhere in Montana, Wyoming, or Idaho, for that matter.

Just as we get past the last buffalo, Frances says, "Look!" Ahead of us on an exposed flat of river shingle a herd of pronghorns has come down to drink: six does, three fawns, and a buck. The male especially looks unlike any modern-day animal. The face is two-toned—fawn and white—and each of its two black horns, short but strangely prominent, is laterally flattened with a forward-pointing tine. He looks as if he has stepped straight out of the pages of my childhood *How and Why Wonder Book* of prehistoric mammals.

Pronghorns belong to their own unique family, the sole surviving species of Antilocapridae, endemic to North America, and were brought to the attention of science by Lewis and Clark. They are called antelope by the locals, but are much less antelope than bison are buffalo. (America does, in fact, have a native antelope, but it is called the Rocky Mountain *goat*.)

Greg French

I mention to Frances that, because of my childhood reading, the pronghorn is the mammal I most wanted to see in Yellowstone, and having a herd of them so close to us seems fair compensation for having to abort our fishing session.

"We would've had to leave anyway," she says reasonably. "I don't really want to have to walk into Pebble Creek in the dark, not in bear country."

"You're right," I concede. "I keep forgetting where we are." At home, since my teenage years, I have had a tendency to walk in and out of the wilderness by night, mainly not to waste any of the weekend's daylight hours, but also because in summer the days can be punishingly hot. It's not as silly as it often sounds to the uninitiated; on the moors, there are only a few nights each month when it's so dark that you need to use a torch.

I have walked long distances at night in the British Isles too, and in Iceland and on continental Europe. I've even done it in Mongolia. It's a strange land, America—more strange to me than any other place I've fished.

PEBBLE CREEK

We park alongside the Northeast Entrance Road at the Warm Creek trailhead or, in parkspeak, at 3K4. There are only two other cars, a pickup displaying Park Service logos and a sedan with a permit on the windscreen telling us that the owners have permission to collect invertebrates. "Government employees one and all," I say to Frances. "Looks like we'll have the fishing to ourselves."

It's three o'clock. Parts of the big blue sky have turned dense gray, especially around the mountaintops, but the air remains warm. Isolated thunderstorms, said the forecast. We don our packs, but before we begin our hike—to the closest tent site on Pebble Creek's headwater meadow, 3P5—Frances diligently attaches a canister of bear spray to one of her front straps and tests a quick draw.

Bob and Karin offered us the mace, or capsicum spray, as we walked out their front door, which was a mild relief to me because Frances had insisted that we take time out to buy some before leaving Bozeman. (And believe me, the word *canister* seriously underplays the device's military proportions. No woman could conceivably conceal this in her purse.)

The walk is a small one, climbing just a thousand feet over two miles, all on formed track through conifer woods, mostly Engelmann spruce and lodgepole pine.

Given the opportunity, I'd prefer to be hiking into the upper reaches of Slough Creek, a famous fly-fishing destination. We had e-mailed our Slough Creek camping application prior to the April 1 deadline, when early requests for backcountry reservations are opened at random and allotted on a first-opened, first-served basis. We were unsuccessful. We can still go to the Slough but only on a day trip, since they are not subject to quotas or bookings.

Most campsites in Yellowstone are not so heavily booked. Getting our preferred camp at Cache Creek was no problem, nor our preferred camp at Heart Lake. We late-booked our camp at Pebble a couple of weeks ago and that hadn't been a problem either, ditto an extra day at our Heart Lake campsite. I suspect that Slough Creek is so popular these days because it is the most respected cutthroat stream in the park. In all probability it has absorbed a few of the people displaced from Buffalo Ford and the more popular tributaries of Yellowstone Lake. I say *a few* because the camping quota doesn't allow for any more than that.

The Overland Track in Tasmania caters to some 12,500 back-packers each year, Routeburn in New Zealand 11,000. On these routes each campsite might reasonably accommodate twenty or thirty tents. The seven campsites scattered along almost twelve miles of Slough Creek pretty much equate to seven tents. Technically, each site can accommodate up to eight people in three tents, but only if a family or club makes a group booking. In practice, most people camp as couples or trios. Similar low-quota regulations are policed along all other walks in the park too.

Greg French

Just a few hundred yards into the forest we hear thunderclaps, and within fifteen minutes rain is falling steadily. The canopy is not completely closed and the drops are implausibly fat. We stop to put on rain jackets, and a couple heading out stops to chat. The man is wiry and middle-aged with a huge gray moustache. He is carrying a fine-meshed net, like a windsock on the end of a broom handle. I point to it and say, "Bit big for aquatic nymphs."

"Butterflies," he says. "We're legal and all. Did you see our permit on our windscreen?" He is proud of his permit.

The woman, more portly and no less friendly, explains that they are writing a guidebook on Yellowstone butterflies.

"Discovered any new or rare species?" asks Frances.

"Last year we found a Pine White," says the woman. "Not previously recorded from the park."

"A real coup," concurs the man.

The two butterfly buffs talk louder and louder about their project, to be heard over the increasing rain, and by the time we part company I am left in no doubt that the book will be comprehensive and authoritative. It's a funny field, lepidopterology: even in Tasmania our most respected experts have had no formal university training, just passion and a professional way of recording their observations. Many fly fishers are the same with mayflies and caddis-flies.

The rain has stopped by the time we cross the ridge and begin our short descent to the Pebble Creek meadow. The creek is small with a gravelly bed, lined with spruce and pine. The water is clear and riffly, mostly ankle deep. There has been a recent flood, and gravel has washed out onto the forest floor.

We cross the creek and enter the meadow. The ground displays an absurd abundance of rodent holes, and has been thoroughly tilled. It also displays an absurd abundance of flowers, and shimmers with purples and yellows. Natural or not, the meadow looks exactly like a ploughed Australian paddock infested with weeds. No doubt about it, what we know about a place influences what we perceive as beauty. Why do I find tailing heaps around abandoned mines offensive but moraines at the front of retreating glaciers wonderful?

We walk a few hundred yards downstream and quickly find our tent site nestled in a copse of conifers right on the riverbank.

So here we are: less than an hour from the main road in the world's most-visited national park in the peak summer holiday period. Completely, utterly, delightfully alone.

It is early evening now, and Frances volunteers to cook dinner while I explore the creek.

Mathews and Molinero claim that Pebble Creek was originally called White Pebble Creek, and I like the way the name's construction parallels that of the Yellowstone River.

Mathews and Molinero also say, "Fishing can be good here for cutthroats that average 10 to 14 inches." Bob and Karin agreed, saying that when they fished it a few years ago they caught several trout bigger than fourteen inches, all from undercut banks on river bends. So I am hopeful of catching a few fish from one to one and a half pounds. Maybe a two-pounder.

At home a fastwater creek like this would produce good numbers of tiny wild browns, and I am disappointed that after an hour of prospecting with a dry fly, then a nymph dropper, I have seen precisely nothing. The problem seems to be that there is no holding water. It's as if a mountain of gravel has washed down through the valley and filled all the pools and undercuts. I stare closely at every tiny depression and see only a couple of fish, both sprats. Again I wonder if cutthroats truly are the pushovers they are reputed to be.

I recall reading an article about a small sockeye salmon creek in Alaska. Originally this spawning ground incorporated a series of holding pools, but after a spectacular flood the pools filled with freestone and the waterway became one long continuous riffle, so shallow that the humpbacked male salmon had nowhere to rest up without exposing half their bodies to the air. Easy pickings for bears. Researchers manning a fish trap noted that not a single male salmon made it all the way upstream, and grave fears were held for the future of the spawning run. But four years later a substantial cohort of salmon returned from the sea. It comprised ordinary females along with males with virtually no hump. Apparently there had always been a small number of no-humped males within the local salmon population, and these fish had been the only ones that survived to pass on their genes. Evolution in the blink of an eye. Revolution.

I am walking back into camp now and it's almost dark. I see Frances ahead holding her index finger vertically in front of her lips. Two large rodents sit like meerkats ten yards in front of her. "Marmots?" I whisper.

"Ground squirrels," she whispers back, holding up our Yellowstone field guide.

At home it would be possums or quolls.

It's nice sitting around the fire, and I don't mind eating vegetables and rice, but I miss the fresh meat and aromatic spices we carry into the bush at home. Bloody bears. At least the Cutthroat Blend is a fine drop.

There is no bear pole at this camp—presumably bears aren't deemed to be a problem here—but before we retire, Frances gets me to put the food in a pack well away from the tent. These sorts of precautions are alien to me. In Tasmania, we keep our food inside the tent to protect it from possums; in New Zealand, we do the same to protect it from keas (cheeky alpine parrots) and hedgehogs. The only care I take is in Tasmania's Western Lakes wilderness, where I lay my fishing rod on chest-high shrubbery so that devils aren't tempted to crunch the cork handle (they are attracted to traces of fish slime).

As we drift off to sleep we hear bloodlust: a wolf pack closing in on its prey. The action is happening at the foothills on the far side of the valley, only a few hundred yards away. The howling stops abruptly, and I figure that whatever the pack had been hunting was small.

I toss and turn all night in my tent beside Pebble Creek. When I wake, the forest is quiet. We make breakfast. Frances watches the ground squirrels. I fish alone. Without success.

The walk out of Pebble Creek is uneventful, and once we reach the car it's time for lunch. We drive a few hundred yards down the road and pull into a classic American picnic area on the banks of Soda Butte Creek, also a tributary of the Lamar. This upper section of the Soda Butte is much bigger than Pebble Creek, but small compared to how big it is when it exits the Ice Box Canyon near the Pebble Creek confluence. No one is fishing.

The tables are shaded by old-growth pines and there are public vault toilets nearby. There are quite a few people around, but you wouldn't know if you didn't check; our nook, like all others, is remarkably private.

I find myself increasingly impatient to begin the walk into the mid-reaches of the Lamar Valley. The Lamar trailhead is only twelve miles back down the road, and the trail, almost dead flat, traverses just four miles of expansive grassland. Yes, tonight we will be camped at the confluence of Cache Creek and the Lamar River: three days of serious fishing on serious water. I can't wait.

SODA BUTTE CREEK

I am holding a billycan, about to scoop water from the Soda Butte for our picnic tea, and I'm looking at the first nonnative fish I have seen in Yellowstone National Park. It's right at my feet in ankle-deep current, shaded by a grassy undercut. An eight-inch brook trout. I hate it for being here. I could catch it and kill it, but I don't want to do that: beauty is beauty, a life is a life. Anyway, as a population-control measure it would be as useless as swatting flies.

In my early years of trout fishing in Tasmania I developed a fascination for brook trout, though I realize now that it was only because of their local rarity. The species had been first introduced to my island in the 1880s, but they failed to establish in the wild, and the hatchery stock was allowed to expire. Brook trout were reintroduced in 1962, the year of my birth, from the Cobequid Fish Culture Station in Nova Scotia, Canada, and a hatchery population has been maintained ever since. In 1963, some fingerlings were released into Clarence Lagoon, a highland tarn that contained no other trout by virtue of the fact that it was located above a small barrier fall. There was a critical short-supply of spawning gravel in the tributary creeks, but a fickle self-sustaining population somehow managed to subsist. What was especially attractive about this lake was its remote location, requiring a two-and-a-half-mile walk or rugged drive over an unformed track. To catch a brook trout, you had to really want to catch one.

When I first fished Clarence Lagoon, in the late 1970s, the brookies averaged two or three pounds and sometimes attained eight pounds. Still, the sport was not especially good. The fish were difficult to spot because of the tea-colored water and mottled substrate, they didn't tail in the marshes, and risers were generally scarce or nonexistent. But the brookies' beauty was impossible to deny, and because of the effort required, each trout landed was cherished.

I ended up tracking down other wild populations of brook trout in New Zealand. In the Hinemaiaia dam on the North Island, I polaroided dense schools of sleek two-pounders that moved in formation like flocks of pigeons. At Lake Emily on the South Island, I caught fish in excess of four pounds. In headwater creeks in the Otago high country, I found stunted fish that coexisted with, or perhaps outcompeted, wild brown trout (a situation unprecedented in Tasmania).

The novelty value of brook trout endured in Tasmania mainly because the species was so hard to cultivate. The government had only one hatchery—the original Salmon Ponds on the banks of the Plenty River, established in the 1860s—and because of the subsequent removal of riparian vegetation and the increasing extraction of water for irrigation, the Plenty's water had become too warm for brookies. Summertime temperatures were particularly problematic, adversely affecting immature eggs as they developed in the brood stock's ovaries.

This changed in the late 1990s when, as an insurance measure, some of the Salmon Pond's brood stock was given to two private hatcheries with the understanding that the government would be given a proportion of the offspring. For the first time in Tasmania, improved water quality and fish husbandry enabled the successful cultivation of large numbers of brook trout fingerlings. The Inland Fisheries Service began stocking these fish willy-nilly throughout the state, and the novelty value of the wild Clarence fish was greatly undermined. Not only did I lose interest in them, so too did everyone else. In any case, the hatchery "catchables" were always outcompeted by wild brown trout and rarely grew to decent size, let alone maturity. They were hardly catchable at all.

I have maintained an interest in fly fishing for brook trout, but only for wild fish, and mainly for wild fish in native environments. I wonder, would I be more interested in hatchery-reared stock if it was my only option, if I couldn't afford the luxury of travel?

The brook trout in Soda Butte Creek are wild but not native, not even long established. According to the National Park Service's *Frontcountry Fishery Inventories*, they have existed for decades in the headwaters outside the park boundary—the source population being a small, unnamed tributary upstream of the McClaren Mine near Cooke City—but were not found in the park until 2003.

The reasons for the sudden expansion of the brook trout's range in the Soda Butte system are unknown, but the spread of the species understandably worries local biologists and anglers, many of whom quote Behnke:

"The biggest threat to cutthroat trout is the introduction of non-native trout, especially rainbow trout, with which the cutthroat trout hybridize, but also brown trout, which replace native cutthroat trout in larger streams; brook trout which have commonly replaced cutthroat trout in small streams; and lake trout, which replace cutthroat in large lakes."

In 2004 and 2005, local authorities poisoned the source population. Also in 2004, they began an ongoing program of electrofishing

Greg French

the upper reaches of Soda Butte Creek. This program included its tributaries from Cooke City to the Ice Box Canyon (some six miles downstream of our Warm Creek picnic area), releasing the natives and killing the brookies. This has been somewhat successful—the numbers of brook trout have declined dramatically in both absolute and relative terms—but electrofishing offers no hope of eradication and the program will need to continue in perpetuity. In my mind, and more importantly in my heart, Soda Butte is no longer a wild ecosystem but a managed one, even a contrived one.

Driving down the valley, we notice where Soda Butte Creek suddenly plunges into the Ice Box Canyon and disappears from sight. The rim of the canyon is level with the road; from the car it looks as if you could step over it. We stop driving and go for a walk. On closer inspection the width of the canyon varies from about five to ten yards. We try to assess the depth, but access to the lip is so steep and gravelly that we chicken out. It is said, however, that because the bottom never sees sunlight, ice remains on the walls long after the rest of the park has thawed. I've also been told you can't fish the Ice Box because the stiff current has scoured the substrate back to bedrock and it is too slippery to wade.

The canyon looks to be half a mile long, and then Soda Butte emerges as a perfect trout stream, like the upper section, only bigger and more fishy. The road keeps tracking alongside the true right bank until we reach the head of the Lamar River Trail.

Quite a few vehicles line the pullout, but I'm guessing that most belong to day trippers, and that most day trippers are anglers. From inside the car we can't quite see the water, only the trail traversing the grassland flanking the south bank. There are one or two walkers, but no one carries a full-sized backpack. We get out of the car and look over the embankment. A dozen or more fly fishers are scattered along half a mile of stream. The water here looks like a mini version of the Lamar proper. It's much clearer, though, and I think it funny that no one is polaroiding from the banks. Not even the people wading the currents seem to be polaroiding. There are no rises either; everyone is prospecting.

One angler directly below us hooks up, and I can tell that it's a fine cutthroat of about two pounds.

Frances and I place our backcountry permit on the dashboard, don our packs, and walk fifty yards or so to the footbridge that marks the beginning of our hike. A couple of fly fishers are resting on the riverbank. They give us a friendly wave, and we stop for a chat.

They are a husband and wife, both in their sixties, from New York. They tell us that these days they can afford to work less than they used to, and that they station themselves in Bozeman for the entire summer so they can fish Montana and the park.

"How's the fishing been today?" I ask.

"Good, very good. And we're not finished yet, I can tell you. The secret is, you have to drift your fly hard, hard in along the undercut banks. Not many folk keep their flies in the zone, at least not when the fish aren't rising and giving their positions away."

"So on a day like today, when the rise is slow, do you go for a dry or nymph?"

"Nymphing?" says the man with exaggerated disgust. "That's not really fly fishing, is it?"

The question is meant to be rhetorical, so I ask if they like to spot the fish before making a cast.

"No need for that."

"What are you catching?"

"Cutthroats. Most are twelve to fifteen inches, some eighteen inches."

"No brookies this far downstream?"

"Not that I've ever seen, and my wife and I have been fishing here every summer for three decades."

"People didn't used to see brookies upstream either," I point out.

"No, but above the Ice Box the Soda Butte is a small mountain creek, and of late, summer flows have been goddamn terrible. The slow current up there has been good for brookies this last decade, no doubt about that, but the current's always been too stiff for them down here. I imagine that when flows return to normal, the brookies won't do so well up there."

"Unless the drought has let them get a toehold," his wife cautions. Then she looks at us. "Where are you going with those big packs?"

"Cache Creek."

"Never been there," she concedes. "Always sort of wanted to, but the fishing is always so good right here that it's hard to make yourself go farther afield."

We wish them well, they us, and we head off.

We can see for miles over the grasslands. There's no shade, but clouds are starting to form and it looks like another thunderstorm might be arriving soon. After a few hundred yards, I venture down to the riverbank. I can't polaroid anything in the riffles, but when I look into the shaded undercuts I can see the fish that the couple were referring to. The conditions for spotting are perfect, and I am tempted to get in and wade. But I see a few anglers working their way methodically upstream toward me. They won't catch the fish I'm looking at—they are drifting their flies down the centerline of the river and on the outer edge of the main seam when they should be hard in along the undercuts. Still, it would be rude to jump in front of them, so Frances and I continue our hike. We stop to talk to a couple returning from a day hike to Specimen Ridge, and they are the last people we see on the trail.

CACHE CREEK

Immediately after setting up camp, Frances and I wander through a hundred yards or so of pine forest and find ourselves on the banks of the Lamar a few hundred yards upstream of the Cache Creek junction. The water is still discolored, but there are long esplanades of open scree and it's pleasant enough working our way upstream, prospecting the main currents with big hopper patterns. Pretty soon Frances calls out that there are some fish rising on the far bank, and at the same moment I notice a good fish rising midstream a short cast in front of me. It clomps my hopper, and is bigger than I expect. It's colorful too—a classic yellow cutthroat—and I call to Frances for help with some photography. We are busily clicking away and changing lenses when Frances suddenly clutches my arm. Her voice is very soft, very slow, very deliberate.

"B.E.A.R."

I look up. Less than ten yards away, slowly foraging its way toward us, is a huge grizzly. I am about to ask Frances why she spelled out the warning when the bear lifts its head and he and I make eye contact.

The surreal calm that has enveloped me for the past two seconds bursts like an aneurysm. Blood pounds in my ears.

The bear puts his head down again.

Frances and I are carefully backing away when my line goes tight. Shit! The fish is still attached. I give the rod a sharp tug, trying to dislodge the fly, and mercifully out it comes.

The fish begins to drift groggily downstream. We retreat to a high bank, and have a perfect view as the bear calmly swims out to the fish and devours it with a single crunch and swallow. The bear then swims to the far bank, craps, walks a little way downriver, swims back across, and starts walking toward us again.

Rather than retreat through closed woods to camp, we choose to wade across the Lamar, walk down along the open banks, and then wade back up the shallow lacework of currents in Cache Creek.

On the way we meet two Alaskans, a father and son out for a day trip, and we mention the bear. "Yeah, we saw him."

"He doesn't bother you?"

"Well, we get to see a lot of grizzlies at home. Leave them alone, they leave you alone."

At the junction of the Lamar River and Cache Creek, fish begin rising furiously to tiny PMDs (pale morning duns), and I learn that there's nothing like catching dozens of pound-plus cuts to take your mind off a bear. We even manage to land a few fish in excess of two pounds.

Evening comes on and we build a big campfire from dry flood debris. The fact that we are allowed to do this seems incongruous with the way the rest of the walking experience is so heavily regulated. In Tasmania you are not allowed campfires in any of our World Heritage national parks—not that locals pay much attention to such petty regulations, not in remote areas where there is plenty of limb-fall and no one to offend.

By now the bear incident seems more of an adventure than a threat. We enjoy the chatter of red squirrels and the flavor of red wine, and as darkness falls, we lie on our backs looking up at the heavens, memorizing the Northern starscape.

Here, in the Northern Hemisphere, I have no innate sense of direction. Day is bad enough, what with the sun tracking across the southern part of the sky, but at least I can *calculate* time and my position in space. By night, I am lost.

I have brought a Northern Hemisphere star chart with me and am determined to become intimate with the local night sky, even if I find it as hard as learning a new language.

The first thing I want to be able to identify is the pole star, or North Star. Strangely enough, my chart doesn't specify it. I've been told that the star is found at the tip of the Little Dipper's handle. I look hard into the heavens. No success. Apparently the other way to find it is to locate the last two stars on the Big Dipper; they are supposed to point to the North Star. There, that must be the Big Dipper! And yes, there's the Little Dipper. Finally I am able to identify some Northern constellations: the Big Dipper, part of Ursa Major (the Big Bear) and the Little Dipper, Ursa Minor (the Little Bear).

Bloody bears.

After dinner, Frances and I wander down to the creek and wash every last skerrick of fishiness from our hands. We take off our fishing clothes too, and put them in our backpacks along with all our food and other odorous items, like toothpaste. Then we hoist our kit up the bear pole—a horizontal log bolted across two tree trunks, fifteen feet above the ground. Finally we retire to our tent and quickly fall into relaxed sleep.

I awake with a start. It's the goblin hours. I hear the chuckle of Cache Creek.

I listen harder.

Yes, something did wake me. A large animal. Padding around, sniffing inquisitively, mere yards from my head. My heart pounds stupidly. I realize I am scared. Really scared.

I try to fight off this panic. I have always found it absurd, even unbecoming, the way some people never come to terms with their own mortality. Every morning since I can remember, I have reminded myself that my days are numbered, and I have regularly asked myself what I would do differently with my life if I learned that I had only one year left to live. I have always made the appropriate adjustments to my life, and in this way death has seemed less dangerous to me.

In truth, though, I suppose I didn't really come to terms with death until my late teens and early twenties when I began spending ever more time rafting Tasmania's wild rivers. These were undocumented drifts on small but steep streams, all littered with unexpected falls, rapids, and logjams. More than once I found myself trapped beneath the water, lungs bursting, seriously believing that death was unavoidable and imminent. And in those moments, everything became dreamy and peaceful. Death, I realized then, was nothing to fear.

So why am I so shit-scared right now? The fear itself, rather than the prospect of dying, is deeply disturbing. I try to pull myself together, to gain some perspective. I've never had to endure life in the trenches at Fromelles as an Australian World War I soldier, not like some of my forebears; I have no right to be scared. I think of all the people who took the Wild West in their stride—Lewis and Clark, Custer's soldiers, Crazy Horse's braves, the prospectors who went to the California gold fields—but my fear doesn't abate; it rises to almost unbearable proportions. I consider waking Frances, but what good will that do?

Several long, sleepless hours go by. The animal sometimes moves away, sometimes goes silent. I begin to understand that this is the sort of fear that sends people mad. Had I served in the Fromelles and survived, I would definitely have returned home shell-shocked.

At the crack of dawn the animal comes back. Not sniffing anymore, but snorting. My nerves are completely shot. I wake Frances and say quietly, "The bear's outside. Put on some clothes. I'm going to sneak a look."

Very cautiously I unzip the tent. Frances grips her bear spray like it's a German hand grenade. On the other side of Cache Creek we see a lone buffalo snorting his way downstream. "You sure it was a bear?" says Frances.

Okay, the snorting was clearly not the bear, but the padding sounds around the tent must have been. I grew up in the country and know the sound hooves make against river cobbles. The sound last night was padding, not clopping. And the breathing was... I don't know, it was just different than that made by cows and bulls.

"I *think* I'm sure it was a bear."

We crawl out of the tent and move away from the trees on to the open shingle flats where we have a clear view of everything around us. We don't see any bear. We calm down. We light a campfire and eat breakfast. Frances, catching a glimpse of the lone buffalo crossing the river upstream, goes for a closer look. She returns after a few seconds and says that she can't see where the buffalo has gone. This doesn't feel right to either of us; buffalo are not secretive. We wander cautiously up the valley looking for the buffalo, and suddenly, lying on the path, we see the bloodied spine of a mule deer or small elk.

Frances says, "That wasn't there yesterday when we walked in."

We break camp in record time, cross Cache Creek, and look back. There's the bear, snooping around our freshly doused campfire.

"He's a good bear," Frances says. "Not aggressive, works at keeping his distance."

Maybe so, but we're still happy to be leaving. We call into camp-site 3L1, a few hundred yards away from our own, to warn the oc-cupants, but there are no occupants. Despite the bear incident, it still seems strange to me that the Yellowstone backcountry is so underutilized.

The other thing that surprises me is that the sun doesn't burn. In our haste we forgot to lather up with sunscreen, and after almost an hour of walking under a cloudless sky we realize we are not red, or even mildly pink. No wonder American visitors to Tasmania forget to protect themselves. Frances, an occupational therapist who specializes in burns, is sometimes called to deal with the aftermath. Thank goodness the world got its act together and reduced the use of fluorocarbons in order to protect the ozone layer. Why can't it do the same thing about carbon dioxide in order to protect the climate?

We meet a group of three anglers, Montanans out for a day trip. We ask where they are headed, but they don't give much away, not even when we mention the bear. They merely wish us a nice day and head off to do what they were always going to do.

Maybe we have overreacted to the bear, the way Kiwi anglers overreact to Tasmanian snakes, or European surfers overreact to Tasmanian sharks. I'm beginning to feel a bit silly.

YELLOWSTONE'S BUFFALO

Near the junction of the Specimen Ridge Trail, we pass through a dale that is clearly used as a wallow by one or more herds of buffalo. We consider walking down to the Lamar in order to salvage the afternoon's trouting, but we have no desire to fish while wearing fully loaded backpacks, and we are not supposed to leave packs unattended unless they have been hoisted up a bear pole, something supplied only at the formal campsites. We take our packs back to the car, then return, a detour of about three miles.

I soon land a good fish, a classic Yellowstone cutthroat with golden flanks and underbelly. I release it, and remembering the way the buffalo coalesced around me when I first fished the Lamar, I take stock.

There are fewer buffalo this far upstream, but nonetheless we are being watched by several lone bulls and a few small groups of cows with calves.

When humans arrived in America 16,000 years or so ago, buffalo were much bigger animals than they are today. In fact, the ancestral bison was so different from the modern buffalo that the two are considered to be different species (*Bison antiquus* versus *B. bison*). About 14,000 years ago the planet began to warm up, people began to flourish, and the megafauna began to die out. By 10,000 years ago North America had lost all of its giant mammoths, stag-moose, ground sloths, giant beavers, camels, saber-toothed cats, maned lions, and dire wolves. But the ancient bison became increasingly abundant and started to herd together in unprecedented numbers. By 5,000 years ago it had evolved into the "small" modern buffalo I see around me.

No one really knows why the bison prospered and changed, but it is widely speculated that the expansion was in response to humans and the new grasslands that humans, with their fires, helped create. In any case, by the time Europeans arrived in the 1500s, bison numbers had grown to an estimated 40 million. The Europeans hated every last one. They hated them for damaging their fences and crops, they hated them for occupying land that might be better utilized by domestic cattle, but most of all they hated them for being *wild*.

Even though Yellowstone was covered in ice until 14,000 years ago, bison have had a long history here as well, one dating back at least 10,000 years.

No one knows for sure what the population was like immediately prior to the park's creation in 1872. At this time human occupation, though traditionally considered too low to affect the "natural" number of animals, was historically high, being bolstered

Greg French

by the westward march of displaced Indians. And the Indians now had horses, which made hunting easier than it had been for their forebears. Also, their buffalo-hide tepees had become bigger, and waste had become common.

In the 1870s, there were reports of several hundred buffalo in the Lamar Valley, and a handful of people pleaded for their protection. In the 1880s, while arguing the case for the construction of a railroad through the park, Representative Lewis E. Payson of Illinois thundered, "I cannot understand the sentiment which favors the retention of a few buffaloes to the development of mining interests..." He seemed to be supported by the weight of public opinion, and throughout the 1890s poaching was endemic.

By this time the few animals left in the park were so persecuted that when finally killed, they often displayed healed or partially healed bullet wounds. They were described as being extraordinarily muscular and flighty—very unlike the fat, lazy buffalo that had once filled the Great Plains.

Just over the ridge from where I'm fishing is the remote Pelican Valley, the place where in 1902 the last wild buffalo made their last stand. It is shocking to think that a population of more than 40 million wild buffalo could be reduced to just twenty-three individuals.

The National Park Service felt that this tiny remnant herd was too small to survive and imported twenty-one domestic buffalo from private ranches, some of which may have been contaminated with the genes of domestic cattle. Still, the fact that the buffalo around me are *essentially* wild is what I like most about them. I certainly wouldn't have come to Yellowstone to see milking cows.

Of course, even the term *essentially wild* is relative, and genetics represent only part of the problem. From the outset the northern herd was fed hay during winter, mainly because the animals found the going tough and preferred to migrate to lower altitudes if they could. Then there were concerns that the herd was overgrazing the native grasslands of the Lamar, so numbers were culled to meet a nominated carrying capacity, a figure that varied from a few hundred to a thousand according to the whim of whoever was responsible for the herd at the time.

By the 1950s, many biologists and naturalists were openly wondering if the herd could really be considered wild if humans had to constantly intervene to make sure the Yellowstone ecosystem remained viable. On the other hand, they also knew that Yellowstone was not a discrete biosphere; "natural" processes couldn't take place. In harsh winters, the Lamar buffalo tried to follow their historic migration route down the Yellowstone Valley, and many exited the park at Gardiner, where they were promptly, and legally, shot. Still, managers and the general public were tending toward the idea of letting nature do most of the deciding.

In 1966, after the Yellowstone herd had been actively reduced to fewer than 300 animals, culling was stopped. It was expected that the natural carrying capacity of the park would see the herd stabilize at about a thousand animals. In recent years, even after the reintroduction of wolves in 1995, the population has often been in excess of 4,000 animals.

The legal slaughter of the buffalo outside the park's boundaries, which continues to this very day, stems from a deal cut by the Interagency Bison Management Plan. This agreement allows buffalo to be hunted, slaughtered, or relocated if they leave Yellowstone Park because the livestock industry insists that buffalo can transmit brucellosis (*Brucella abortus*) to livestock. There is some speculation that the disease may have arrived in the park in 1903 via a domestic cow that was used to suckle two wild buffalo calves captured in the Pelican Valley. On the other hand, park managers have from the outset speculated that brucellosis might be native to Yellowstone, and if so they would be obliged to protect the bacterium as an essential part of the ecosystem.

Apparently this disease can cause livestock to abort their calves and produce undulant fever in humans, and infected animals cannot be transported to other states. So the Yellowstone buffalo are confined like indigenous people on reservations.

73

Greg French

It annoys me that buffalo are persecuted. I find it especially galling because actual cases of brucellosis transmission have never been verified. Though maybe I shouldn't be so cavalier, not when I know too well how supposedly benign fish diseases have suddenly and unexpectedly wrought so much havoc around the globe, and right here in Yellowstone too. But the crux of my animosity is this: I don't think brucellosis is so much a problem as an excuse. Rural people have always hated buffalo. Elk also carry brucellosis, and although they are well known to have transmitted the disease to domestic cattle, they are not persecuted. The only difference, I suspect, is that the elk are a major recreational resource for local hunters. Rural people have always revered elk.

There is no doubt that the Yellowstone buffalo represent a conservation success story, but there are myriad uncertainties about what to do next. Is the buffalo population unnaturally high? If so, why? Do the buffalo overgraze the Lamar Valley? Should the buffalo be prevented from damaging delicate thermal features?

If maintaining biodiversity is the aim, we have to intervene. If allowing natural processes a free rein is the aim, we have to decide what "natural" is and what we are prepared to sacrifice.

And what if these questions are beside the point? What if human interests inevitably lie at the core of everything we do? What if we conservationists are really arguing about the identification and preservation of certain vital but contradictory aspects of human spirituality?

RAINBOW TROUT

I am still thinking about all this—the morality and passion that underlies concepts like "changed by human agency," "wild," "natural," "nonnative," "genetically pure"—when I realize that my fly has disappeared from the surface. I strike... and all hell breaks loose.

Eventually, the fish is lying on its side in the shallows, exhausted after a long fight. It is a much bigger fish than my first one, brilliant silver with a just hint of pink—a rainbow. The contrast between the two trout could not be more striking. *Oro y plata*. Both are precious and beautiful: you'd need a touchstone to tell which was the most valuable. Or if either was pure.

In *Trout Fishing*, Joe Brooks wondered if the rainbow trout in the Yellowstone system (all found downstream of Yellowstone Lake below the Lower Falls) might include a natural population derived from individuals that had, in prehistoric times, migrated across Two Ocean Pass. Modern DNA testing, in addition to geographical evidence, tells us that this is not the case: all the rainbows in the Lamar are derived from translocated stock.

Here again the question of natural purity rears its head. In pre-Colonial times, coastal cutthroat trout and rainbow trout shared a common range and occasionally interbred. And despite the fact that the chromosome difference is 68 for coastal cutthroat and 58 to 60 for rainbow trout, the hybrids were fully fertile. So why aren't cutthroat trout considered merely a subspecies of rainbow trout? The truth is, biologists have yet to come up with any ironclad way of defining what a species actually is. There might not be any plausible way to do so.

I first became aware of this at age eleven when reading *Trout Fishing*, in which Brooks described several varieties of brown trout native to Ireland's Lough Melvin. The gillaroo, he said, was "physiologically different" from other trout, having a hardened stomach resembling a gizzard that enabled the fish to cope with a diet that consisted predominantly of sharp-pointed snails. The sonaghen, on the other hand, was small (three-quarters to one and a half pounds) and "behaviorally different" from other brown trout, going to sea only once in its lifetime and thereafter schooling at depth in the lake (but still happy to rise through twenty feet of water to take a dry fly). Brooks also mentioned the ferox trout, which was said to

Greg French

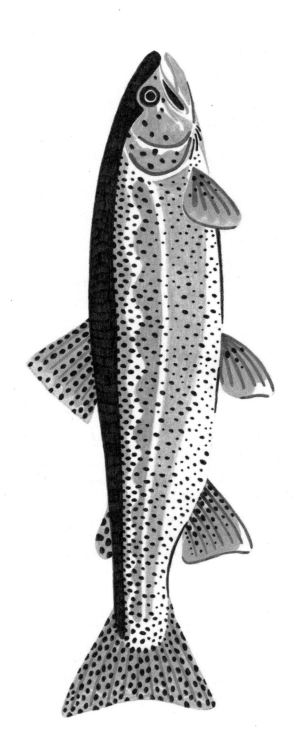

Rainbow Trout
(*Oncorhynchus mykiss*)

live deep and feed almost exclusively on other fish, including tiny landlocked char.

Everyone knew that the three types had coexisted for thousands of years without losing their separate identities, but in the 1970s the consensus amongst biologists was that they were all the same species and that the differences between them were the result of random behavioral and environmental influences. Recent DNA analysis, however, has thrown the neat arrangements into turmoil. In *About Trout* (2007), Behnke noted:

> "The old-timers were right: the gillaroo, sonaghen, and ferox of Lough Melvin are real entities, not figments of an active imagination or due to something they ate.... They do not interbreed with each other.... The implications of such situations go far beyond simplistic taxonomic arguments over lumping versus splitting. Populations with different life histories and ecologies should be managed as separate species. The problem, or gray area, for classification is the fact that, among salmonid fishes, the occurrence of closely related populations living together in the same lake is such a common phenomenon that if we recognized every such population as a separate species, names would proliferate and classification would become chaotic."

But clearly chaos is the natural order of things, and clearly it is disingenuous to ignore complexity simply to preserve a neat model for classification. On the other hand, Behnke makes a valid point: if we dilute our innate understanding of *species* too far, the term will no longer carry any practical value, or even an intrinsic one.

The concept of *species*, vague as it is, somehow remains vitally important to us anglers. We want a rational reason for sticking to our guns about cutthroat trout being fundamentally different from rainbows. The best science can come up with is this: despite the fact that the rainbow/cutthroat hybrids are fertile, strong natural selection used to keep them from spawning successfully—no self-sustaining populations of naturally occurring wild hybrids have ever been documented. And for whatever reason, that definition is good enough for me: it gives me something to treasure. But goddamned hatcheries have turned this paradigm on its head.

The problem of hatchery fish breaking down barriers to hybridization is so pervasive that it now happens even where species have historically coexisted naturally. But the most stunning rate of hybridization occurs when rainbow trout are introduced into cutthroat waters above barrier falls, or over the Continental Divide.

Where rainbows come into contact with inland subspecies of cutthroat trout, as opposed to the coastal variety, rampant interbreeding can occur almost instantly; Colorado's yellowfin cutthroat trout (*Oncorhynchus clarki macdonaldi*) was apparently hybridized to extinction within twelve years of hatchery-reared rainbow trout being introduced to Twin Lakes in about 1890. Other times, rampant hybridization is delayed, but no less devastating when it occurs.

Here in the Lamar system, native Yellowstone cutthroats and introduced rainbows have coexisted as separate species (albeit with some hybridization) for a century or more, yet in recent years the barriers to hybridization, whatever they may have been, have started to break down at an increasingly rapid rate. It seems that past resistance to interbreeding is no guarantee that the integrity of stocks can be maintained in the long term.

These days the rainbow trout in the Lamar system are reviled, and the extermination of the species is presented as a moral imperative, a fight of Good against Evil. But what if the demonization of the rainbow trout is merely a human imperative? After all, the Yellowstone cutthroat trout themselves don't give a damn about the rainbows. They are quite happy to mate with the enemy, and the offspring of such unions are healthy and robust, as robust as "humans" like me who carry Neanderthal genes.

As for the things trout eat—the green drakes, pale morning duns, blue-winged olives, golden stones—things haven't really changed for them over the last century or so: to get eaten by a cut or a bow is equally abhorrent.

I strongly suspect that a hundred years from now the argument for preservation of genetic purity as a moral imperative will be viewed as quaint. Even now, in our hearts, we know it is the lesser part of a more important story. What of the domestic buffalo liberated among the last wild buffalo in 1902—genetic pollution? Certainly. But it is almost universally agreed that the Yellowstone herd retains some sort of vital integrity. Yellowstone, you see, is the only place in North America where a "wild" population has persisted since prehistoric times. It's not genetics so much as history and context that's so important here. Human values.

In the early 1900s I would have revered my Lamar rainbow trout, reveled in the difficulty of the catch, and treasured the added variety. Today, my Lamar rainbow is troubling, but only because I desperately want the Yellowstone cutthroat trout to survive. I won't resort to defending my needs on "moral" grounds, or even practical ones. They are entirely spiritual. In the modern world, the homogenization of culture and Nature is all-pervasive. These days, difference comes at a premium, and the cost is counted in moral uncertainty.

I wonder if the biggest single difference between today and yesteryear—the real reason most anglers could so easily live without rainbow trout in the Yellowstone catchment—is that travel has never been so cheap or convenient. These days if we want to sample exotic fish, we can so easily travel to the places where they "truly belong."

Chapter 4:
Slough Creek

ABSAROKA

We were supposed to spend a second night at Cache Creek, but since that's no longer on our agenda, where are we going to sleep? A little luxury would be nice, but if we are going to find a room anywhere at all, it won't be within the park. Months ago, when planning this trip, we had tried to book accommodation in Canyon Village, Tower-Roosevelt, and Mammoth, but everything had been booked a year or more beforehand. We had to settle for a room at the Absaroka Lodge in Gardiner. That booking is for tomorrow night, but maybe we will be able to extend our stay backward by one day.

Most people pronounce *Absaroka* ab.suh.**rock**.uh, though those who live within a couple of hours' drive of the mountains tend to say ab.**zor**.kuh. We have even heard some say ab.**zor**.kee. Apparently, the Indians themselves prefer ab.**suh**.uh.kuh. Take your pick. The name means something akin to *children of the big-beaked bird*, which the French interpreted as "people of the crows," hence the popular English term I grew up with: the Crow Indians.

At the time of European arrival in the Americas, the Absaroka were the nomadic people of the Yellowstone Valley, roving from the foothills near present-day Bozeman (Montana) to the Missouri junction (on the border of present-day Montana and North Dakota). They were constantly harassed by larger tribes like the Sioux and Cheyenne, and were driven farther and farther west. By the time of the Battle of the Little Bighorn, they were largely confined to the area between the Bighorn Mountains and the Absaroka Range.

Here they had become noted horse breeders, and were constantly being raided by tribes that did not have as many horses, or horses of such high quality.

We drive into the Absaroka Lodge's bitumen car park. I look at the garish seventies architecture, and my heart sinks. At reception we learn that there is one room available, and my heart lifts.

The room is mundane. Still, it's better than the alternative, I suppose. We unpack our dirty camping gear and get ready to go to the Laundromat. While I wait for Frances to shower, I go out to the balcony and lean on the handrail. The views of the Yellowstone River are superb. Within the confines of the room my eye lights on the ubiquitous TV, provided in case the views aren't sufficient. As we are staying in the Absaroka Lodge, I start thinking about *The Thundering Herd,* the film based on Zane Grey's book, and then rewind back through an embarrassingly long memory of westerns.

The first movie I remember watching the whole way through until the credits was broadcast on our brand-new black-and-white telly in 1967 when I was five. *The Last Hunt* was a western, a real western with Indians and buffalo. I was on their side, instinctively understanding the Indians' relationship with the land. The scenes of buffalo being shot by the hundreds, the carcasses being left to rot on the prairies, the sadness of the native people ... all this affected me more than I can say.

Throughout *The Last Hunt,* the sepia photo-portrait of General Custer in my encyclopedia kept galloping though my thoughts. I remember asking the adults around me if the Indians might have loved other animals too, trout perhaps, and I remember the polite scoffing.

Tasmania's own indigenous people were not mentioned in my encyclopedias, nor by my primary school teachers, and perhaps this helps explain why I became so fascinated with the lore of the Indians instead of the lore of my own island. I remember wondering if I had any right to recruit as my own the passion of a people so

Greg French

very far away in space and time. Yet I felt some sort of affinity, even ownership, of American Indian spirituality:

> "The earth does not belong to man, man belongs to the earth. All things are connected like blood that unites us all. Man did not weave the web of life, he is merely a strand in it. Whatever he does to the web, he does to himself."

For years I thought that these words belonged to the Plains Indians—the Sioux or Absaroka. Not until adolescence did I find out it was attributed to Chief Seattle, a Squamish Indian from Puget Sound in Washington. Not until middle age did I find out that even this attribution is disputed. But, as Bill Thomas once wrote in *FlyLife* magazine, "It doesn't matter a fig whether this is a direct quote or not. It's not a bad sentiment."

The history of the Plains Indians remains as fascinating to me now as ever, and before coming to Yellowstone, I reread my child-hood encyclopedias and Googled as much as I could.

Historically, it seems, the Sioux were made up of three great divisions: the Eastern Dakota (from east of the Dakotas), Western Dakota (from the Minnesota River area), and Lakota (the western-most Sioux of the Great Plains).

By 1700, some Sioux had moved to what is now South Dakota. In about 1730, they were introduced to horses by the Cheyenne, and thereafter their society centered around hunting buffalo on horseback. As a child, after several years of watching westerns, my realization that the horse was not native to the Americas shocked and disappointed me. At first I refused to believe it; I thought it might invalidate the hunt. I tried to define *natural* and *unnatural*, and nothing I came up with gave me any comfort. I really struggled to work out what was legitimate and what was not.

The Lakota's initial contact with United States officialdom occurred during the Lewis and Clark Expedition (1804–1806). The natives immediately distrusted the whites, and so they should have.

In 1851, the Lakota and the United States signed the Fort Laramie Treaty. The Indians guaranteed free passage to travelers on the Oregon Trail, and in return the United States agreed that the Lakota would retain sovereignty over the Great Plains "for as long as the river flows and the eagle flies."

I loved the Indians' sentiment. My brother thought that the Indians were stupid. "If the river stopped flowing and eagle stopped flying, they would lose everything to the United States. You should always keep something up your sleeve," he said. Like Custer, he didn't get it.

In 1874, the Sioux and Cheyenne got their first dose of Custer, who was sent to "examine" the disturbances in the Black Hills area. He was on the side of the white invaders. So too was General Philip Sheridan, who encouraged his troops to slaughter buffalo in order to destroy the Indians' commissary, military jargon for food supplies and livelihood. But by the time I was a teenager, I understood too well that food and livelihood were the least of what was lost. I knew by then that if you took away my trout, what you would really do is break my spirit.

When the United States government proposed that the Northern Pacific Railroad should be routed right through the middle of the reserved Sioux hunting grounds, the deceit and dishonor became too much for Sitting Bull and Crazy Horse.

In primary school I learned that Sitting Bull once said, "If the Great Spirit had desired me to be a white man he would have made me so in the first place. He put in your heart certain wishes and plans, and in my heart he put other and different desires. It is not necessary for eagles to be crows."

It matters not a fig how accurate this saying is; it fitted me as a child like no other saying ever did.

The crossed musket insignia that so scared me as a preschooler—mainly because it recalled pirate crossbones—was adopted by the United States in 1875, and the next year the army was formally at war with the Sioux.

The Great Sioux War, which raged from 1876 to 1877, was really a series of guerrilla strikes that included the Battle of Powder River, the Battle of the Rosebud, and the Battle of the Little Bighorn. The US Army, I was saddened to learn, was supported by the Shoshone, who apparently believed the maxim, *The enemy of my enemy is my friend.*

Inevitably, perhaps, the United States military emerged victorious. The Lakota were forced onto reservations where there were no buffalo, where they had to rely on the United States government for food.

I read about all this in the early 1970s, at the same time as I was reading Brooks's *Trout Fishing*. At the same time my coastal hometown happened to be building a giant woodchip mill, a giant seaweed processing plant, and a giant fishmeal factory. I tell you, I was on the Indians' side.

In my despair, I thought more and more about the Indians' horses, and I devised a thought experiment:

What if the *buffalo* were not native to North America? What if the Indians had brought them across the land bridge from Asia? How would that affect the legitimacy of the hunt? Of the buffalo themselves?

The ramifications of my questions complicated my world and filled me with dread.

And now, here in the Absaroka Lodge, it occurs to me that the concept of "legitimacy" remains as elusive to me now as it was when I was a youngster. Even though my childhood fear that the modern buffalo were introduced by humans turned out to be baseless, there is no doubt in my mind that the evolution of the ancient bison into the modern buffalo came about as the result of human agency. In wilderness areas, changes brought about by "human agency" are supposed to be unacceptable. So exactly how much change is unacceptable? And why?

I'd like to shelve these thoughts and relax with a beer, but tomorrow Frances and I will fish Slough Creek, where questions about the limits of acceptable change cannot be sidestepped.

SLOUGH CREEK

We get up early and have breakfast in Gardiner. We pay for our food, and among the change there's a commemorative quarter. According to the waiter there are fifty different designs in total, one for each state, minted from 1998 to 2008 in the order that the states ratified the constitution. If our kids were young, we would have fun collecting the coins. As it is, we have fun checking for coins we haven't seen before. I especially like Kansas (a buffalo), North Dakota (two buffalo), Washington (a leaping salmon), Alaska (a grizzly bear catching a salmon), and Minnesota (a loon, with people in the background fishing from a boat). Montana opted for a buffalo skull, while Wyoming opted for the same silhouette of a rodeo rider that appears on its car registration plates. It's a bit disappointing that no one chose a trout. Some states actually preferred to depict commerce. At least no one stooped as low as our local council did back home. Our emblem is supposed to be a hop kiln and ream of newspaper, reflecting the major local industries, but it looks exactly like an outhouse and a roll of toilet tissue.

We drive back into the park toward Slough Creek. We've been playing the registration plate game for a few days now, and new states are ever harder to come by. Frances spots Alabama, and I'm miffed. I make a distance call on another plate, but Frances points out that it's just Montana, that different designs from the same state don't count. She actually accuses me of cheating.

We turn off the Northeast Entrance Road. After two and a half miles this gravel spur will terminate at a campground, which the Park Service describes as primitive. I guess that's because generators are not allowed, but they supply hand pumps for drinking water and vault toilets. Tent sites are available on a first-come, first-serve basis, but everyone tells us that if you don't arrive before 8:00 a.m., you aren't in the running. That's why we didn't come here yesterday afternoon.

85

Greg French

The last two-thirds of the road runs alongside the lower reaches of Slough Creek. It's very good water flanked by flat grasslands, and quite popular with fly fishers. Mathews and Molinero say that the broadwaters are relatively deep and produce superb hatches. "Native Yellowstone cutthroats have the edge in numbers, but non-native rainbows and hybrids dominate in size."

We are more interested in the backcountry, however, where until recently no rainbows penetrated at all. Exactly why the rainbows have started moving up past the campsite through the gorge into the upper meadows no one is prepared to say, though I'm starting to think that low water resulting from a decade of drought probably has something to do with it. In any case I'm happy to be able to sample the fishing before "genetic pollution" really takes hold. Many other people feel the same way; the purity of the fish is one of the reasons that the upper section has always been as popular as the more accessible lower section. As Behnke noted in *Trout and Salmon*, "Many anglers find great joy and satisfaction in the pursuit of fish which are native to the region where they are fishing."

The first meadow is located an hour's walk upstream from the trailhead. According to YellowstoneNationalPark.com, the catch rate here is about five trout per hour, and the fish average fifteen inches (which I figure is one to one and a half pounds, depending upon condition).

The second meadow is two hours from the trailhead. There is less pressure here and the trout are supposed to be far more readily caught. YellowstoneNationalPark.com suggests that this area is second only to the Yellowstone River at Buffalo Ford in terms of angler success. We will probably start our fishing here, but if there are too many anglers I am quite prepared to walk further up the valley.

The third meadow is three hours from the trailhead, and by all accounts the fishing is essentially the same as that in the second meadow. It is not often used by day trippers, presumably because most locals and visitors are daunted by the minimum fifteen-mile return hike. Among my peers back home, however, such a walk is not that extraordinary. We have all had days polaroiding lakeshores in Tasmania's Western Lakes wilderness where we've notched up fifteen or more miles in a day, ditto in New Zealand's backcountry streams, even on the South Island's giant braided rivers.

There are lots of cars at the trailhead, but they might just be overflow from the campground or even belong to anglers fishing downstream. As for people, there are only three: a guide and his two clients, both of African descent. In the South Pacific most dark-skinned fly fishers are islanders, but these guys tell me that they are Americans from the Florida Keys, where they love hunting the flats for bonefish, tarpon, trevally. They are the first dead-keen polaroiders I've seen in America, and I know that they will take every advantage of the day's opportunities. We wish each other well.

Frances suggests that we give them a bit of time to go on ahead, so we boil a cup of tea. I wipe the sweat from my brow. The day seems to be following the usual pattern: it'll attain seventy-five degrees Fahrenheit by midday, maybe into the eighties in the afternoon, and sometime between two and four, there'll be a thunderstorm. Things will calm down by the evening, and if we come home after dark, we will probably find the temperature dropping to the high fifties.

The first part of the walk skirts inland to avoid the gorge, which turns out to be a short section of granite country containing a few sets of small rapids.

Even though it would have been nice to have secured a back-country camping permit, we're glad to be walking without a full pack. We stroll through conifers and cottonwoods and see count-less red squirrels and chipmunks, even get up close to a couple of yellow-bellied marmots. Ah, this is more like the North America I remember from British Columbia.

In a remarkably short time we break out into the first meadow. The track stays a bit south of the slow, serpentine stream, but we can see all of both banks. The only people are the guys we met at the trail-head. The Floridians are well in front of their guide, running along the banks to get casts at fish they are polaroiding. Good on them.

A lone bull buffalo ambles along the track several hundred yards ahead, and Frances and I think it prudent not to overtake him. There's ample room for many more anglers in the first meadow, and I'm tempted to fall in behind the Floridians, but instead we use the time to take photographs.

After ten minutes we don our daypacks again, and we soon cross over the ridge and gain sight of the second meadow. Impatient to start fishing, we immediately strike off toward the river, and end up in a section of riffles bounded by granite outcrops and sparse pines, where we quickly polaroid a few small fish and land a couple of two-pounders.

Forty minutes later, we arrive at the meadow proper. We can easily polaroid the silty bottom of the first broadwater, but we don't see anything. So I look hard, hard in along the edge where a stiff current undercuts a grassy bank. I'm almost sure I can see a few tiny, tiny sipping rises. I make an exploratory cast with a big deer-hair hopper, and suddenly I'm into a feisty three-pounder. In the next forty minutes we catch a dozen solid fish, and I say to Frances, "Well, I'm bored now, let's go home." I'm exaggerating, of course, but there's no doubt that I'm becoming less obsessive about the day's fishing than I've ever become in Tasmania's Western Lakes or on New Zealand's backcountry rivers.

The strangest thing is the frogs. They are big—three inches long—and they're constantly jumping out of our way from the grassy banks into the river, yet none attracts the attention of a trout. I try an experiment whereby I throw a live frog in front of a rising fish, producing no result, and then cast a grasshopper imitation in front of the same fish, and the fly is eaten in a flash. From the field guide we work out that they might be northern leopard frogs (*Rana pipiens*), but we can't be sure. Is it a case of co-evolution resulting in frogs that don't taste nice to trout? That would be truly weird considering the berserk feeding frenzies undertaken by wild brown trout in Tasmanian marshes.

It's only mid-afternoon, so we decide to walk back via McBride Lake, which I'm told has a fine population of native cutthroats, most of which weigh a bit less than one pound but some of which grow to more than two pounds. There is no trail, so we follow a small gutter up through pines and granite outcrops, yelling, "Yo, bear!" all the way.

The lake turns out to be a green soup, with dense clouds of granulated algae blown hard in along the shore. Some of the bays are actually viscous. Is this normal?

The water quality is a real pity: if not for the algae, the elevated banks would be ideally suited to spotting, and there are plenty of fishy-looking wind lanes and drop-offs. We walk to the far end of the lake, then strike off downhill back to the first meadow.

It's late afternoon by now, but not too late, and we are surprised to have the meadow entirely to ourselves. I catch some small fish where the riffles enter the meadow, and quickly move downstream. There's a reasonable current in the broadwater, but the surface is flat-calm and I have a perfect view of the silty substrate. I polaroid a few two-pounders midstream and catch them. Then all the fish move hard in along the banks to mop up spent spinners—tiny pale morning duns (PMDs). The rising trout are not in the least bit spooky, but they are almost impossible to catch. I go small. I try matching the hatch. Nothing works, not consistently. The fishing becomes so captivating that we don't notice how late it is until we hear the howling of wolves. We really should leave before dark, we suppose, what with the bears and all.

I don't know how many fish we've landed for the day, but I am relieved that none were obvious rainbows or hybrids.

The howling continues as we walk home. Frances and I smile at each other. Even more than the buffalo, the wolves are one of Yellowstone's great success stories.

People's historical treatment of wolves, in practice and in literature, reflects a primal fear similar to the one we still have for snakes

Greg French

and sharks. Probably this fear was exacerbated when people became farmers and wolves started attacking their precious livestock.

What spelled doom for the wolves of America's Wild West was the era of railway construction in the 1860s and 1870s. After the great majority of buffalo were killed off, the starving wolves turned their attention to domestic cattle, and the settlers let loose with firearms, leg traps, and poison.

Even the newly created Yellowstone National Park was subjected to an enthusiastic wolf-control program whereby random carcasses were laced with strychnine. According to the park's website, from 1914 to 1926 about 140 wolves were killed by park rangers alone. After that there were no more wolves to kill. In fact, by this time the wolf, formerly the most widely distributed mammal in the United States, had disappeared from almost all states, and the only ones that remained were living near the Canadian border.

In the 1930s, there was a surge in sentiment among biologists and naturalists for preserving the "total ecology," that is, environments that included natural predators. The situation didn't really change, though; in the early 1960s Steinbeck wrote that the killing of wolves, coyotes, and similar "vermin" was still seen as a patriotic enterprise. The real turning point didn't come until 1962, when Rachel Carson published her highly influential book *Silent Spring*.

In 1972, after years of intense lobbying, President Nixon finally banned the poisoning of predators, and in 1973 (the year I read *Trout Fishing*) Congress passed the Endangered Species Act. One of its mandates was to restore endangered wildlife to native habitats, and so began the push to re-establish wolves in Yellowstone National Park.

From the viewpoint of a conservation-minded Tasmanian, the ferocity of opposition to this proposal seems ludicrous. But hunters and self-proclaimed patriots were very, very angry; ranchers were incensed.

Conservationists combated the opposition with a slick public relations campaign, but the battle wasn't won until the ranchers were guaranteed the right to kill wolves that strayed from the park.

Finally in 1995, thirty-one animals of breeding age were released into the wild, and within a decade the species had become a major tourist attraction. At least fifteen packs were using the park in 2006, and goodness knows how many there are now.

The reintroduction of wolves has had some very surprising follow-on effects: why on Earth should they cause beavers to bounce back from the brink of local extinction? The answer turned out to be stranger than fiction.

When they had no predators, huge herds of elk would stay for prolonged periods on the riverbanks, where they would chew saplings—willows and cottonwoods—to ground level. Nowadays the elk are constantly harried by wolves and forced to stay on the move. The saplings—many of which are inadvertently planted by the beavers—are able to flourish, and for the first time in decades the beavers have adequate food and building materials.

The impact of the wolves on beaver populations makes me wonder what effect saber-toothed cats, mastodons, and dire wolves had on the environment before people wiped them out. It also makes me wonder if a park without such keystone species can really be considered to be a "wilderness largely uninfluenced by human agency."

In any case, Slough Creek feels so wild and exciting to us that we arrive back at our car before we are quite ready to stop walking. Oh well, there's plenty more of the park to explore tomorrow.

Greg French

Chapter 5: Mammoth to Norris

MAMMOTH AGAIN

We vacate our room at the Absaroka Lodge early in the morning and drive to the Mammoth backcountry office. I'm a bit annoyed that we weren't allowed to pick up our Heart Lake permit when we picked up the permits for Pebble Creek and Cache Creek.

The young ranger has grown more confident since we saw him last. "How were your hikes?"

"Loved them," Frances replies. "Had a bit of a bear experience at Cache Creek, though." She's trying to sound casual.

"Lucky you," he says. "You know, plenty of experienced local hikers have never seen a bear. How close?"

"Less than ten yards, and *coming right for us*," says Frances, amusing the ranger with her exaggerated *South Park* accent.

But despite his laughter, the ranger is concerned, more so when we tell him about the bear hanging around the campsite. "Grizzly or black?"

"Grizzly."

"You sure?"

"We can show you photos if you want."

He nods and I pass him my camera. "Oh my God, that's a class-five bear. A carcass, you say?" He hands us a flipbook of dead deer in various stages of decomposition, ranging from a fresh kill to a pile of bleached bones. We pick one halfway between, with lots of jerky-like meat on it. "Oh my God, a class-three carcass." Suddenly he is on the phone to the bear response group, and within minutes the lower Cache Creek area is closed to camping.

The ranger explains that these days it's unusual for bears to become accustomed to campsites. When they do, they become problem bears and potentially dangerous. "A SWAT team will go in straight away, maybe by helicopter. They'll use aversion therapy: set up small scented beanbags of mild explosives in the campsite, even take potshots at the bear with rubber bullets if need be."

"Do we have to watch *Beyond Road's End* again?" I ask with an exaggerated groan.

The ranger laughs. "You think you've got it tough. Spare a thought for my ranger buddy. One of his ex-girlfriends is the presenter—acrimonious breakup—and some days he has to sit through this film ten times."

We get our permit for Heart Lake in record time, and walk out of the dingy office into bright sunshine. The elk are still everywhere on the lawns, and a ranger is politely warning visitors not to get too close. We ask him why there are no stags.

"Stags are loners. They don't want nothing to do with no one or nothing until the females come in heat, sometime in August. You'll see a stag or two on your way to Norris, if you're going that way. They already have full racks of antlers, a real fine thing to see, and you'll know when one's close because there'll be a queue of cars pulled over on the road shoulder. You have a nice day now."

Our plan is to fish Buffalo Ford and spend the night at Yellowstone Lake. We've already made an advance booking at the car-accessible Bridge Bay camping complex.

I hate advance booking anything—spontaneity is part of my being—and it's been an anathema to me that Yellowstone requires such detailed forward planning, such rigid scheduling. As an act of rebellion, I say to Frances, "Let's look at the hot springs before we leave Mammoth."

She smiles quizzically, knowing full well that I don't have much interest in thermal features, much less overhyped tourist attractions.

"At least they aren't on our itinerary," I say in self-defense.

Adjacent to the main road, hundreds of people line the wide

timber boardwalks that give access to various limestone terraces, hot springs, spring cones, communities of brightly colored heat-tolerant bacteria, and stands of scalded trees. As we join the crowds, I wonder why nothing I see really incites my passion. Sure, it's all interesting—beautiful, even dramatic—but I feel none of the ecstasy I feel when fishing, see nothing that will lurk siren-like in my dreams and woo me back.

Strangely enough, a geology buff picks me out as being especially interested in the landscape and attempts to strike up a conversation. I reply with less than the expected enthusiasm, and realizing that I have hurt her feelings, I try to atone. "It's not that I don't like geology. It's just that geothermal areas don't inspire me as much as glaciers and icecaps. Maybe it's because my home state, Tasmania, has been so obviously sculptured by ice. Maybe I appreciate the distinct association between ice-cold water and trout."

"You're a fly fisherman like my boyfriend, aren't you?" she deduces brightly. "I'm studying limnology. Volcanic activity can affect waterways and aquatic animals just as much as glaciation does, you know."

She's right, of course. I mention a fishing trip to the Tongariro River and Lake Taupo on the North Island of New Zealand in 1995 when I saw the eruption of Mount Ruapehu.

"Wow! You've actually seen a volcano blow its top?" she says. "The only chance us Americans have had was in Washington State in 1980 when Mount St. Helens went off. I wasn't even born yet."

It seems incongruous to me that most people in Montana and Wyoming have never seen an eruption, considering that Yellowstone National Park is pretty much one gigantic volcano.

"Me and my boyfriend went to Taupo last year," the girl continues. "I was surprised at how big the lake was—bigger even than Yellowstone Lake."

I ask her if she fishes.

"Only a bit. My boyfriend is way into it. Most men are in Montana. Wait." She walks to the other side of the boardwalk and pats a young man on the shoulder. He turns around, and she leads him back to me. "This is Joe."

"Greg," I say, offering my hand.

"And I'm Rachael," the girl says. "Greg's a fisherman from Tasmania," she explains to her boyfriend. "He fishes the Tongariro and Lake Taupo."

Joe and I end up in deep conversation about fishing on the North Island, and while comparing notes it dawns on me that the aftermath of the Ruapehu eruption is still dramatically apparent at the fishery two decades on.

During the year-long period of major volcanic volatility, ash killed most or all of the fish in many of the major spawning streams. Then, upon being flushed into Taupo itself, the ash altered the chemical nature of the lake, resulting in a boom of surface-dwelling plankton. Suddenly trout that normally retreated to the thermocline during summer remained visible all over the lake surface. And because of the superabundance of food, the fish began piling on weight. In recent seasons, however, most of the fish have reverted to old habits. Yet some things have not returned to normal. The size and condition of the fish are now historically low. And the once-huge autumn run of rainbow trout spawners—one of the main attractions for anglers fishing the Tongariro—is in dire straits.

I wonder: *How much silt and ancient volcanic ash flowed into Yellowstone Lake following the catastrophic wildfires of 1988? What effect did it have on the lake's water chemistry and the cutthroat trout fishery?*

MADISON RIVER

Rather than travel back through Tower-Roosevelt, it makes sense to take the alternative route through the northwestern quadrant of Yellowstone National Park to Norris, especially now we've been told about the likelihood of seeing elk stags. And by going this way,

I figure I might get the chance to fish for some native westslope cutthroat trout.

I didn't originally plan on fishing for westslopes in Yellowstone —I'd caught plenty during our trip to Canada in 2006—so I don't really know where to go. The guidebooks aren't much of a help because they only talk about cutthroats generically, and Wi-Fi reception has been hopeless since we left Gardiner. I end up leafing through some of the pages I printed from the park's website before leaving Bob and Karin's place, including an article titled *Restoring Fluvial Populations of Native Trout.*

I learn that within the park, the westslope cutthroat is native only to the north-flowing Gallatin and Madison drainages, and most populations have been displaced or outcompeted by nonnative trout. The only remaining native population of genetically pure fish was discovered in 2005 in Last Chance Creek (deliberately named), a tiny tributary of the Grayling Creek in the Gallatin headwaters. Other genetically pure populations exist in the similarly tiny Oxbow and Geode creeks (both in the Yellowstone drainage between Mammoth and Tower-Roosevelt), but these are the result of artificial stocking in 1922.

I also learn about the High Lake Westslope Cutthroat Trout Restoration Project, which, in 2006, involved poisoning nonnative Yellowstone cutthroat trout in tiny East Fork Specimen Creek.

Frances says, "How can it be that nonnative cutthroat in the Oxbow and Geode are highly valued by conservationists while nonnative cutthroat in East Fork Specimen Creek are despised?"

"Well, it's obvious, isn't it?"

"If you think it's obvious, you don't understand the question," she replies.

As was expected, the poison used in East Fork Specimen Creek—rotenone—not only killed all the fish but pretty much all other aquatic animals, including native insects. Yet the carnage was deemed acceptable because the nonnative cutthroats would be replaced with "native" cutthroats (even if the seed stock did not have any ancestral link to East Fork Specimen Creek). Moreover, the aquatic insects would re-colonize naturally.

Personally I am happy with the "restoration" project—it does not diminish fishing opportunities and it does increase variety. I am troubled, however, by the rationale. Those responsible for the program present their actions as a moral imperative, a shift in management style from the anthropocentric (human-centered, utility-based) to the eco-centric (environment-centered, wilderness-based). Yet surely the primary purpose—perhaps the only purpose—is to preserve the spiritual needs of wilderness lovers like me.

I suppose that the ecosystem will have regenerated by now; westslope cutthroats will have been reintroduced and insects will have re-established. In fact, the state of affairs probably is more or less exactly as it was before: the westslope cutthroat trout will probably exist in exactly the same size and numbers as the Yellowstone cutthroat trout before them, and they will probably eat exactly the same quantities of exactly the same prey.

Still, can the project in any meaningful way have benefitted wilderness when wilderness is not sentient? One thing's certain: the poisoning definitely didn't benefit the wild animals it killed.

If humans are the sole beneficiaries of the restoration project, that's fine by me. I would only be uncomfortable if humans had to intervene on a regular basis, if the streams were not subsequently left to Nature, to serendipity.

Curious to learn more about the westslope cutthroats in Yellowstone National Park, I consult Behnke's *Trout and Salmon* and Anders Halverson's *An Entirely Synthetic Fish*.

Apparently, from prehistoric times to the late 1800s, the westslope cutthroat trout flourished throughout the main stem and most of the major and minor tributaries of the Madison River. The only other native salmonids were the Arctic grayling (*Thymallus arcticus*) and mountain whitefish (*Prosopium williamsoni*). In the late 1800s, rainbow trout and brown trout were introduced, presumably for variety or better sport, and by 1915 the westslope cutthroat and Arctic

grayling were all but extinct in the upper river; not much later they disappeared from most of the rest of the system.

But the demise of the native salmonids would prove to be only one of several hatchery-induced catastrophes to affect the Madison fishery.

By the 1930s, the extraordinarily robust populations of non-native brown trout and especially nonnative rainbow trout in the Madison's "Fifty Mile Riffle" (from the Hebgen Dam just outside the park boundary downstream to Lake Ennis) had become famous among local and international anglers. And by the 1950s, the Riffle was attracting tens of thousands of anglers annually. To support this pressure, Montana's Department of Fish, Wildlife & Parks began stocking the river with hatchery rainbows, most of which had been reared to catchable size. Every year thereafter, the locals complained about diminishing returns. Yet increasing the stocking never seemed to solve the problem.

In 1966, Montana finally employed an official fish counter—a new graduate in fisheries biology from Montana State University, Dick Vincent—whose job, in part, was to figure out why the Madison River seemed to be spiraling into terminal decline. He eventually demonstrated that the stocked fish were directly impacting the wild fish. When stocking was stopped, "... the population of wild rainbows would explode by up to 800 percent; the number of wild browns would easily double. And not only that, the wild fish were bigger, more fun to catch, and tastier to eat."

It turns out that hatchery fish always harass wild fish and interrupt feeding and spawning behaviors. Wild fish are displaced, and the hatchery fish quickly die through angling, predation, or the inability to feed properly. The fish that survive interbreed and these hybrids are ill-suited to life in the wild.

Nowadays no one in Montana would consider turning back the clock by prioritizing hatchery fish over wild fish. Yet the poisonous legacy of the hatchery era is far from over. A third hatchery-induced disaster—whirling disease—has everyone in a spin.

Greg French

ALIENS

In 2006, when we were in the British Columbian Rockies fishing for westslope cutthroat, some inland waters had just been subjected to classified water regulations. If you were a British Columbia resident, this was irrelevant; you simply bought an annual classified waters license and continued to fish where and when you liked. But all visitors, including Canadians who lived a couple of miles away in Alberta and understandably considered the East Kootenays to be their own backyard, were forced to pay a hefty tax that was levied per person, per day, per river. The new regulations on the Elk River begged the question, who is an outsider? According to at least one vocal member of the Rod and Gun Club (a local lobby group), an outsider was anyone not born in Fernie. My mate Russ, a local resident, was incensed. "I was born in England, my daughter was born in Fernie. Why shouldn't we be able to fish together? Is it reasonable that we should all be restricted to fishing waters within a couple of miles of our birthplace? Is parochial selfishness a fair basis on which to formulate fishing regulations?" Since then, I'm told, *South Park* has done a piss-take of British Columbia's fear of alien anglers and it's as delightfully ludicrous as the first-ever episode, *Cartman Gets an Anal Probe*.

If aliens are a real threat, it's not the ones from outer space or the ones with fishing rods that are the real worry. In New Zealand, for example, they are invasive species like didymo (a Northern Hemisphere alga). Mind you, which aliens most bother you depends entirely on your personal passions. In 2007, following years of unprecedented drought, Australia's Federal Government created the Murray-Darling Basin Authority, primarily to find ways of guaranteeing a sustainable future for local communities along Australia's longest and most socially important river. It was well known that the main cause of the Murray-Darling's dire ill health was its long and persistent history of grossly excessive water extraction, yet the new authority immediately created a website with a homepage that featured a nonnative trout juxtaposed with a skull and crossbones. To top the list of pressing issues with a nonnative fish seemed plain weird, but if the authority felt compelled to do so, why didn't it use a symbol that was more widely recognized as a major problem? Why not a European carp? Mosquitofish? Redfin perch? Goldfish? Did the website's authors see trout as an *ideological* threat: everyone could agree to hate most nonnative fish—how dare so many people love trout!

The hatred of one or another type of alien is sometimes overcome more easily than anyone might reasonably expect. In 1989, the North Island's Lake Aniwhenua experienced a massive bloom of nonnative water net, a green alga that grows in a mesh-like configuration. This particular alien made polaroiding difficult, constantly fouled flies and lures, choked native waterweeds, and sucked oxygen out of shallow bays. Anglers left the lake in droves. But no sooner had the local council set up an aerial spraying program than local stream enthusiasts began noticing a huge increase in the average size (and number) of fish running up a tributary creek, the Horomanga. It turned out that the water net was an ideal food resource for water fleas and snails, and Aniwhenua trout were gorging themselves before migrating upstream to spawn. The council stopped spraying, and anglers returned in droves. For the next few years, typical fish taken from Aniwhenua and the Horomanga weighed four to ten pounds, and quite a few weighed close to twenty. Then in 1994, the bloom disappeared as quickly and mysteriously as it arrived. By the time I fished the Horomanga in autumn 1995, the fish were a modest three to six pounds and local anglers were trying to figure out ways of triggering another bloom.

So what are the most dreaded aliens here in Yellowstone? Happily, the American Parks Service does not consider people to be among their number; Yellowstone is still proudly maintained "for the benefit and enjoyment of the people," including out-of-staters and international visitors like me. Here the acknowledged baddies are mackinaw, New Zealand mudsnails, and whirling disease. And at Buffalo Ford, I will get to see the havoc they have wrought.

Greg French

Chapter 6: The Yellowstone River

BUFFALO FORD

Shortly after we pass through Norris we see two elk stags, and then we stop for the mandatory peek at the Grand Canyon of the Yellowstone, containing the spectacular Lower and Upper Falls.

According to popular lore, the name *Yellowstone* arises from the river's Minnetaree (Sioux) name—which translates as "Yellow Rock River"—and comes from the color of the walls of the canyon. In truth, the name probably comes from the color of sandstone bluffs much farther downstream, perhaps as far away as North Dakota.

Above the falls, the only native fish in the Yellowstone system are cutthroat trout and longnose dace (*Rhinichthys cataractae*). But there is a long history of alien invasions.

According to Wikipedia, in 1908 some 7,000 Atlantic salmon were stocked into Yellowstone Lake and 5,000 into Duck Lake (which drains into Yellowstone Lake at West Thumb). These fish were sourced from a landlocked strain, but did not establish.

Rainbow trout were liberated too. Ken Retallic's *Flyfisher's Guide to Wyoming* suggests that "a US Fish Commission employee made an unauthorized transplant into Yellowstone Lake in 1907," and notes that more rainbows were released into the upper Yellowstone River in 1929. Still, no rainbows exist above the falls today, nor do the native cutthroat bear any trace of hybridization, either visually or genetically.

In 1985, brook trout were found in Arnica Creek, a tributary of Yellowstone Lake's West Thumb, but the stream was immediately poisoned, then poisoned again in 1986, and the population was effectively obliterated.

According to Wikipedia, mountain whitefish were introduced into the Yellowstone River immediately below the lake but also failed to establish.

Some aliens remain, however. Mackinaw were discovered in the lake in 1994, as everyone hereabouts is acutely aware, and are said to pose a serious threat to the native cutthroats.

The other nonnative fish—lake chub (*Couesius plumbeus*), redside shiner (*Richardsonius balteatus*), and longnose sucker (*Catostomus catostomus*)—do not appear to have had much of an impact on the native fish. In official park literature and various scientific reports, all three are said to have been "most probably introduced by bait fishermen." There is no evidence to support these claims. In Tasmania the carp in lakes Sorell and Crescent were originally said to have been most probably introduced by bait fishermen, but were subsequently found to have been introduced as eggs on the fyke nets of commercial eel harvesters. The introduction of redfin perch into a number of other Tasmanian lakes was also widely blamed on bait fishermen, but it turned out that every one of these lakes had been stocked with domestic rainbows sourced from a commercial fish farm where redfin were endemic.

We are driving past the falls and up the Yellowstone Valley now, and when we get out of the car at Buffalo Ford we will wade upstream as we do in Australia and New Zealand. (We couldn't organize a downstream float because drift boating is prohibited on any part of the Yellowstone River within the national park.)

According to Mathews and Molinero, July 15—"the day the season opens on all fishable parts of the Yellowstone River upstream of the Upper Falls to the outlet of Yellowstone Lake"—is circled in red on fishermen's calendars all over the globe. Apparently the fish that anglers come to catch have recently migrated down from the lake and, after spawning, they take the opportunity to feed heavily amid the huge summertime hatches of riverine insects before returning home.

July 15 has certainly been circled on my calendar. Unfortunately, we have arrived a little later than that—July 27, to be precise. And now I'm starting to doubt that we'll *ever* make it to Buffalo Ford; there are buffalo everywhere, and the traffic is backed up for miles.

Whoop! Whoop! "Now, folks, move away seventy-five feet from all large wildlife. You sir. Yes, thank you. There are a lot of animals going backward and forward across the road, so we ask everyone to please be patient." *Whoop! Whoop!*

We get out of our car and mingle with other people who, mindful of the ranger's requests, are nonetheless taking the chance to stretch their legs and photograph the wildlife. One old man—"Call me Chuck"—tells me he's a local from Cooke City, so I ask him why Buffalo Ford is so named. He replies that it is the spot where his great-uncle and some other rangers found a frozen herd of buffalo. "It was the mid-1940s, just after the Second World War. The buffalo must have forded the river, and the wind chill snap-froze them. They were all looking west toward Nez Perce Creek and the Madison Valley, the way they used to go to get off the mountain in winter, and the tears in their eyes were frozen too. They were probably better off that it happened that way. If they'd made it past the park boundary at West Yellowstone, they'd only have got themselves shot. Cattle ranchers hated them back then. Still do."

I'm not entirely sure of the veracity of Chuck's account. In *To Save the Wild Bison*, Mary Ann Franke tells the story slightly differently: "… thirty-eight bison drowned in January 1946 after breaking through the ice and becoming trapped in the Yellowstone River."

Chuck continues, unfazed. "You know, buffalo weren't the only animals to get themselves frozen in these here mountains. In the fifties, when I was a teenager and another uncle was nearly eighty, he took me for a hike to Grasshopper Glacier. It's north of the park just outside of Cooke City, in the Beartooth Mountains of the Custer National Forest. A long time before we humans turned up, way back when mastodons and woolly rhinos called Yellowstone home, a swarm of locusts got caught in a storm and ended up stuck in the ice. We dug a few dozen of them out with a penknife, and they looked so fresh you could eat them. Which gave us this idea. We took some home and put them in an icebox, and the next week we took them to Buffalo Ford and caught trout on them. Crazy, huh, catching Yellowstone cuts on extinct hoppers? That was my first fishing trip, and Uncle Dave's last."

I already knew about Grasshopper Glacier. A visitor to Yellowstone could scarcely not know about it, appearing as it does in all the guidebooks. The hoppers aren't quite as ancient as Chuck would have me believe, but they are centuries old nonetheless, and the species is indeed extinct. I am reminded of Sitting Bull's vision, immediately prior to the Battle of the Little Bighorn, of soldiers falling into his camp "like grasshoppers from the sky," and I ask Chuck about the best way to reach the glacier.

"Not worth the effort anymore. Last time I went there, there were hardly any hoppers left, and the ones we found were rotten and falling to bits. There's hardly any ice left actually, not compared to when I went there the first time. And you know what, that's why there aren't so many fish at Buffalo Ford anymore."

He senses my bafflement, and it pleases him. "Most winters, there ain't much snow no more," he elaborates.

"Bet the buffalo are pleased about that," I say.

"It's no laughing matter, sonny."

Sonny? Jeez, I'm fifty. "I heard that last year was cold," I press. "Bloody cold."

"Last year was only cold by comparison to all the warm years we've had of late. It sure wasn't as cold as it was in '46. No buffalo froze last year, not so far as I'm aware of. And if you ain't got snow, you ain't got snowmelt, and if you ain't got snowmelt, you ain't got water in the spawning streams. That's the problem with Yellowstone fishing. If those goddamn scientists wanted to help the fish, what they would do is bulldoze away some of those sandbars at the stream mouths that stop the fish getting to the spawning grounds."

"I don't think there's likely to be much support for taking bull-dozers off-road in Yellowstone," I say, trying to sound sympathetic.

Greg French

"But the sandbars aren't natural, sonny," he replies, barely able to disguise his contempt. "They're just ash and erosion from the '88 fires, and there's nothing natural about that, I can tell you. Where have you been these last decades? Don't you know nothing about global warming?"

You wouldn't know it by talking to anglers, but Buffalo Ford isn't called that anymore. In 1981, it was officially renamed Nez Perce Ford, in honor of the Nez Perce Indians. In the 1870s, gold—damned gold—was discovered in Oregon's Wallowa Mountains, the homeland of the Nez Perce, and the United States government insisted that the Indians be moved, by force if necessary. In 1877, pursued by Major Benteen, the angler who one year before had led one of the two battalions that did not get annihilated in the Battle of the Little Bighorn, the Nez Perce fled across Yellowstone National Park. They entered from the west along the Madison River, traveled along what is today called Nez Perce Creek, and crossed the Yellowstone at the spot that is right this minute just half a mile and perhaps several hours'ahead of us. Theirs was a reverse route to sanctuary from the one the buffalo would have followed had they not frozen to death.

Nez Perce Ford is, in fact, the only place where it is possible to ford the Yellowstone River, and if you are a person rather than a buffalo, you can only do it when levels are very low, usually between high summer and late fall. The Nez Perce, with the help of their horses, probably managed the crossing with ease, but were finally defeated in the Bear's Paw Mountains, a thousand miles from home, just forty miles short of safety at the Canadian border.

Given the hardships faced by the Indians, it seems churlish to be grumpy about our forced delay. It's a minor inconvenience really, and anyway the traffic has started moving again.

Hooray! We've finally reached the Buffalo Ford picnic area. Numerous clouds of steam emanate from various sulfur vents and mud volcanoes. Dozens of buffalo crowd the area too, but we manage to park seventy-five feet from the nearest. Frances is a little nervous, but I get out and piece together my rod. I am nervous as well, but only about the fact that there is hardly anyone around. I hope the reason is the buffalo, not the lack of trout.

Mathews and Molinero advise me to "Pick out a single fish, determine what insect it's eating, and match that with an imitation." They strongly caution against "flock shooting."

The first time I saw a flock of fish was when fishing for those wild brook trout in the Hinemaiaia Dam near Lake Taupo. Since then I've seen big flocks of westslope cutthroat trout in the Elk and Wigwam Rivers, and much bigger flocks of lenok in Mongolia's Onon gol, but the descriptions of the Yellowstone fish seem incredible nonetheless.

Edward Ringwood Hewitt, in *A Trout and Salmon Fisherman for Seventy-five Years*, talked of fishing in Yellowstone National Park in 1882, saying that he caught thousands of fish, none of which was bigger than four and a half pounds.

A few months ago I clicked onto YellowstoneNationalPark.com and the "flyfishing" page suggested that this section of river hosts some 4,500 anglers per mile of stream, many of whom catch fifty or more fish per day, the average weight being about two pounds.

Other people say things have changed.

Before leaving Tasmania I e-mailed David Sweet from the East Yellowstone chapter of Trout Unlimited, and he told me that all his angler friends were extremely dissatisfied with catch rates and that fewer people were traveling to Yellowstone to fish the lake or the river. On the positive side, though, he acknowledged that the average size of the fish had increased dramatically, with some now growing more than twenty-five inches, which I figure could be seven pounds or so. He reckoned the weight gain was probably due to the fact that the remaining fish would have to be old ones.

Well, I guess I'll soon see for myself what the pickings are like.

I am about to walk to the river when I notice three returning anglers. I wander over to them and we all lean against the side of their pickup, talking as if we are old mates, the way keen fly fishers are apt to do the world over.

Greg French

"Much going on today?"

"Not a lot. Not much of a hatch. There were some pale morning duns, green drakes, gray drakes, caddis, and even a few salmonflies, but nothing rising. We ended up with three between us, small ones, and then the buffalo moved in. Name's Steve. Where you from?"

"I'm Greg, from Tasmania."

"Australia, right? And we thought we were a long way from home. We're from Utah. This is Rod and Doug. We're old mates. Go on trips like this every year. Buffalo Ford is one of our favorites. It's a lot tougher than it used to be."

We talk some more—about fishing and family and buffalo— and then we shake hands and I wander down to the river. I see that there's good access along a hundred yards of grassy esplanade, but beyond that there's a copse of pines where a number of buffalo are lying in the shade. The water is clear and the bottom gravelly. I can wade about ten yards from the bank, but the current is too strong to go any farther. There are no rises, and I quickly get bored with prospecting, so despite the glare and the tumble of the water, I concentrate on polaroiding. I end up spooking two fish before I see them, both four-pounders in prime condition.

Frances has come down to join me now, so I wade ashore and we decide to try spotting from the higher bits of bank. At the mouth of a small, permanent anabranch is a deep hole, and on the edge of the hole a giant cut is cruising back and forth. It must weigh at least six pounds. We cast all manner of flies to it, but it refuses everything. I move on ahead, up onto a high bank perilously close to the copse where the buffalo are, and spot another big fish. It keeps disappearing into the depths, and I have to wait minutes at a time for it to come back into sight. Just when I feel I can get a good cast in, there is a loud snort of hot air. I look up. A buffalo is lying in a shadow mere yards from where I stand. Other buffalo are moving down the bank toward me.

"Time to go," says Frances.

She's right, of course, but it's a real pity; without the buffalo, this would be champagne fly fishing.

As we pick our way between the animals littered around the picnic area, we walk past one of the few parked cars. An ancient angler leans out the window and says, "You going home, yeah? Can't say I blame you. What's become of the fishing here is enough to make you cry. Way worse than it was in the sixties and early seventies before they introduced catch and release. Goddamn mackinaw."

I wish I could properly see how many fish there are in the river, and how big. If I had a wetsuit and snorkel, I'd be sorely tempted to hop in and drift downstream for a mile or so.

With a face mask under the water, you don't have to fight reflective glare and nothing is obscured by swirl and current. The fish are rarely completely hidden under rocks and usually remain in station until you are almost upon them. You can see deep into undercut banks; everything is as clear as a fishbowl.

Seeing fish is so easy, in fact, that New Zealand researchers frequently use drift-counting methods to estimate fish stocks, and claim to have developed ways to standardize the process and make allowances for variables such as water clarity and current.

I wonder if this is good science. I've been skeptical ever since Frances, Russ, and I met our first professional drift diver on the Elk River in British Columbia.

Russ noticed him first, a few hundred yards behind our boat—a dark shape drifting on the surface. "Can't be a beaver," he quickly deduced. "No wild animal is that uncoordinated. Looks more like a nincompoop in a wetsuit."

Every time we dropped anchor the snorkeler aimed for our boat and managed to get a little closer. Finally we just waited for him to catch up. He lunged at the gunwales and the boat rocked chaotically. He cupped his left armpit over the gunwale, removed his snorkel, pulled back his face mask, and announced that he was Chad, a research officer. Apparently he was evaluating fish stocks by drift diving. And the results were dire—numbers were way down from previous years, stocks reduced to just a dozen or so fish per mile.

How on Earth could one man expect to count all the fish across a river so wide and fast?

Frances was amused. Russ and I were both keen for him to leave. We asked if he wanted us to contribute to his research, and he said no.

Russ told him to enjoy his day, but it still took Chad several minutes to detach himself from the boat.

How surreal. I asked Russ why he didn't contradict Chad's assessment of fish stocks. "What difference would it make?" he replied.

I had to admit, he had a point.

We had counted more fish from a single bankside vantage than Chad saw in several miles. If Chad wanted to monitor fish stocks, he should have been relying on multiple methodologies, including collating the annual catch data from the local guides.

Biological field research often relies heavily on skill and intuition, not to mention the necessity of robust methodology, and is only ever as good as the researchers and their supervisors. And for examples of good research and evaluation, you need look no further than the studies that brought about mandatory catch-and-release regulations in Yellowstone National Park.

CATCH AND RELEASE

In the mid-1970s, the upper Yellowstone provided the first clear proof that if anglers released their catch of wild trout, the average age and size of the fish would increase over time and the catch rate would dramatically improve. This, along with Dick Vincent's findings on the Madison about how much damage hatchery "catchables" could do to a sustainable wild fishery finally led to the winding back of hatchery operations throughout Montana and Wyoming. But why had hatchery operations become ubiquitous in the first place?

The trout hatchery was a European invention—Stephen Ludwig Jacobi is credited with establishing the first facility in Germany in 1741, and the famous Stormontfield Ponds (an Atlantic salmon hatchery) was built on the banks of Scotland's River Tay in 1852—but the idea of trout hatcheries being an imperative was a distinctly American way of looking at things, a way of having your cake and eating it too. You could continue to chop down the forests, dig up the land, channelize the rivers, and the angling could be maintained simply by adding unlimited numbers of hand-reared trout to make up for those that could no longer be naturally spawned. Hatcheries could also replenish healthy streams where overkill was resulting in greatly reduced bags. It seemed too good to be true, and of course it was. Yet for decades the rise of the hatchery seemed to occur in the absence of any real opposition, or any suggestion of alternative management strategies.

Catch and release was also a European invention. Behnke, in *About Trout*, has traced the first record of actual regulation to the book *Salmonia*, written by Sir Humphrey Davy and published in 1928, which made mention of a private hatchery on the River Colne in England where anglers were required to release all trout weighing less than two pounds so that they could be caught again the following year when they would be bigger. But, again, catch and release as an imperative was a distinctly American way of looking at things. As evidence, Behnke, in the essay "We're Putting Them Back Alive," presented a compelling paper trail, beginning in 1902 when David Starr Jordan and Barton W. Evermann published the classic book *American Food and Game Fishes*, which included a heartfelt lament about the blandness of hatchery fish.

More and more American writers began talking about the benefits of releasing fish, and then in 1939, Lee Wulff (designer of New Zealand's favorite fly, the Royal Wulff) published his *Handbook of Freshwater Fishing* in which appear the immortal words: "Game fish are too valuable to be caught only once."

Wulff's slogan was so wonderfully succinct and quotable that as a youth I absorbed it by osmosis and began using it even before I realized it was a quote. And, of course, I have used it more passionately since I became aware of its origins.

Greg French

In the 1950s, Australia's Aubrey Nicholls and America's Paul Needham became two of the first fisheries scientists to strongly prioritize the preservation of habitat and gene pools over the production of hatchery-reared fish as the best way of improving fishing for all.

Taking their lead, the highly respected and influential organization Trout Unlimited was founded in 1959 explicitly to promote the value of wild trout over hatchery-reared imitations, especially over fish that had been reared to catchable size.

In the 1960s, the bulletin of the Sports Fishing Institute (an organization funded by tackle manufacturers) published an article titled "Fishing for Fun," and in 1964, it coined the term "catch and release." Anglers everywhere began to accept that the key to maintaining healthy fisheries was to embrace the preservation and restoration of habitat over the increasingly discredited quick fix of hatchery-enhanced fisheries.

Yellowstone became one of the first major public fisheries to experiment with enforcing catch and release. From 1970 to 1972, anglers fishing the lake were obliged to release all fish less than fourteen inches long, and the bag limit was set at three fish per day. But even with these "severe restrictions," the lake continued to be overexploited. In 1973, the bag limit was reduced to two fish, and the cutthroat population continued to decline.

In 1975, park authorities changed tack: from now on, anglers would be required to release all fish of thirteen inches or bigger, bait fishing would be prohibited altogether, and the bag limit would be maintained at two fish per person.

Finally Yellowstone had stumbled upon a suite of regulations that was actually effective. Incredibly so. By 1978, the spawning run in Clear Creek had increased from less than 10,000 (recorded in the 1960s) to 70,000. By the 1980s, the spawning run in Pelican Creek had increased from 12,000 to 24,000.

There were other benefits too. Before 1975, very few fish reached five years of age, but by the 1980s, fish of seven to eight years old had become quite common and some were attaining nine years. Not only that, catch per angler tripled to two fish per hour. It seemed that the fishery had been saved for posterity.

What wasn't really understood at the time was that the "recycling" of fish would result in improved angling only under certain circumstances. It turns out that the Yellowstone experiment was so tremendously successful only because the Yellowstone cutthroat trout were so easily caught, and because the fish in the lake and the outflowing river were capable of attaining both old age and large size (typically about eighteen to twenty inches or two to three pounds).

When regulations had been scant, the great majority of trout were being cropped off en masse when they were small, before they ever had a chance to spawn. The ease of the slaughter was truly remarkable. In most lakes and rivers, overexploitation occurs when harvest approaches 50 percent of catchable fish per year. A 50 percent harvest of brown trout usually takes at least 500 to 800 hours of effort per surface acre, but Yellowstone Lake and its outflow fishery greatly exceeded this percentage when fishing pressure was a mere five to six hours per surface acre.

Another way of looking at relative catchability is this: cutthroats are about seven times easier to catch than brook trout, eight to fifteen times easier than rainbow trout, and about thirty-five to seventy times easier than brown trout.

FLY FISHING VERSUS EVERYTHING ELSE

I'm in a coffee shop in Grant Village wondering if tweaking the fishing regulations might help reverse the current decline in cutthroat numbers. The rules surrounding catch and release seem pretty robust, so I'm thinking more about the "minor" regulations.

Currently, with the exception of the Gardner River (which has special regulations for children fishing for tiny, nonnative brook trout), you are not allowed to use bait. But spinning is allowed, even on prime fly water, and spin fishers are allowed to use double and treble hooks. The only concession seems to be that all hooks, even trebles, have to be barbless.

Without Wi-Fi reception, I can't Google anything about the successes and failures of minor regulations, so I reread Behnke's article, "We're Putting Them Back Alive."

I am surprised to discover that there is no empirical evidence to support the claim that fish caught on flies are better able to survive than those caught on lures: typical mortality rates range from 0.3 to 3 percent of all fish caught and released, regardless of whether they were hooked on a fly or a lure. Furthermore, an angler-induced mortality rate of 3 percent has no major effect on any trout fishery; it pales into insignificance against natural mortality, which typically runs at 20 to 50 percent of all takeable-size fish per year. And Behnke emphasizes that under normal conditions, virtually all mortality among angler-released fish is caused by hooks puncturing blood vessels in the gills and mouth, not from stress, as is commonly suggested in much popular literature.

The caveat to all this is: mortality increases as water temperature rises above sixty degrees Fahrenheit, and can become very high as temperatures approach seventy degrees. Death in such instances is caused almost entirely by acidosis, though this is not caused by a build-up of lactic acid as is generally assumed, but rather by some unknown acid.

Bait fishing, however, is problematic for those wanting to release fish—even if the water is cold—because peer-reviewed research shows that it results in up to 40 percent mortality. In Yellowstone, where your average fish is likely to be caught 9.7 times a season, bait fishing would result in 999 trout out of 1,000 being killed during the course of a single summer.

Behnke stresses that, from a biological perspective, laws mandating the use of barbless hooks and favoring fly fishing over spinning are a complete waste of time. Worse than that, he's worried that they channel effort and funds away from much more pressing priorities, primarily the preservation of water quality and habitat. I wonder what he would make of the "dry fly only" restrictions on Slovenia's Unica River and England's chalk streams. (I think they're infuriating, especially if you arrive to find all the fish nose-down in the weeds.)

Despite all of this, I concede that in special cases, gear restrictions can greatly improve the angling experience, even greatly popularize it. The best example I can think of is Tasmania's Little Pine Lagoon, a water of such global significance that it was appraised by Joe Brooks in *Trout Fishing*. This small impoundment was formed in the 1950s, and it was apparent from the outset that the spot would appeal mainly to fly fishers, principally because the expansive weedy shallows would make it unsuitable for lure fishing. The worry was that a few lure fishermen would try to fish anyway, and that by barging through the shallows looking for pockets of deep water, they would disturb fish searching for food in the weeds, ruining the sport for the majority. Set-rod fishing was deemed inappropriate too, mainly because lines trailing out into the marshes would restrict access for fly fishers intent on stalking fish.

The lagoon quickly became renowned for two annual highlights—springtime tailers and summertime dun hatches—and developed the highest average catch rate for good-sized wild browns in Tasmania. No matter whether the fishing pressure is calculated in anglers per mile of shoreline or per acre of water surface, Little Pine Lagoon continues to be the most intensively fished lake in Australia. And no one seriously doubts that this is because of the fly-only restriction rather than in spite of it.

So minor regulations can sometimes enhance the social aspects of the fishing experience, but they probably can't do much to help increase the supply of naturally spawned fish. Yellowstone's cutthroat need an altogether different sort of help.

FISHING BRIDGE

We are standing on the Fishing Bridge, which spans the Yellowstone River immediately downstream of Yellowstone Lake. The air is gray and cold, the wind rising. To the west is the Elephant Back, a solidification of lava from the most recent series of volcanic eruptions 70,000 years ago. These eruptions were conventional ones, not ones of the Yellowstone caldera itself, yet if explosions on that scale were to occur today, every single human and trout within the park would be obliterated.

It seems they already are. The huge car parks at either end of the bridge are more or less empty. At first we hoped that the absence of tourists was due to the weather, but after walking the full length of both sides of the bridge we have seen only a dozen or so clusters of spawning cutthroats, none of which included more than a handful of individuals. It's disappointing—during spawning time back home in Tasmania or New Zealand, any major lake tributary would carry thousands of fish.

Oh, for what used to be.

In the late 1800s, before the bridge was built, people rowed to this area from the Lake Hotel—a few miles away across Yellowstone Lake—to witness the unparalleled quantity of spawning trout. In the *Overland Monthly* of June 1897, Frank King talked of catching one-pound fish on flies "two or three at a time" and lamented that it was all too easy. He much preferred fishing for nonnatives in the Firehole.

The original Fishing Bridge was built in 1902. The site was chosen because the water was relatively shallow, but this meant that it was also a major spawning area for Yellowstone cutthroat trout.

It is hardly surprising, therefore, that Yellowstone's new bridge soon became an irresistible attraction for anglers. In *The Outing Magazine* in 1908, Ralph E. Clark mentioned that many anglers killed up to fifty fish in an hour from the bridge, but he too lamented the fish's ease of capture, poor fighting quality, and wormy flesh, and again advised the serious angler to try his hand at catching nonnative fish in rivers like the Firehole and Madison.

The worms of which he spoke were tapeworms (*Diphyllobothrium* species). In Tasmania, tapeworms mainly affect trout in coastal lagoons, where the life cycle of the parasite involves seabird hosts. I wonder what the vector is here, so far inland.

The bridge quickly became known as the Fishing Bridge, and the event helped foster and entrench a huge prejudice against the native Yellowstone cutthroat. In *The Wise Fisherman's Encyclopedia* of 1957, Don Martinez noted:

> "From the standpoint of the casual tourist who is only mildly interested in fishing, Yellowstone Lake and the river are the chief attractions in the park. Trout are caught off Fishing Bridge ... on every conceivable sort of tackle, including cane poles, hand lines, and surf tackle armed with spinners, flies, or worms. A confirmed fisherman will shudder and look the other way—it is vaguely disquieting to see sizable trout hauled out of the water one after the other in plain sight of noisy people."

Modern environmentalists have seized upon this quote, displaying a vaguely disquieting tendency to interpret Martinez's observations as snobbery.

In any case, his account was just about the last of its type. Although no one knew it at the time, harvest had by that time exceeded recruitment. In *Yellowstone Fishes*, Varley and Schullery noted that in the early 1960s some 49,000 anglers used the Fishing Bridge each year, and the average catch rate had dropped to just one fish per angler for every seven hours of effort.

The predicament wasn't resolved until the late 1960s when Jack Anderson was appointed superintendent of the park. Being a serious angler, he made sure to enact regulations to bring the Yellowstone cutthroat trout back from the brink. First, he enforced low-harvest and catch-and-release regulations of the type that were being recommended by fishing societies nationwide, making the park a model for fisheries management worldwide. Then in 1973 he did the unthinkable and closed to fishing the first mile of the Yellowstone River below the lake outlet in order to "protect spawning trout and restore naturally occurring levels of trout in the area."

Modern-day anglers applaud Anderson's efforts because he managed to restore the fishery. Many modern-day environmentalists, however, regard Anderson's efforts as heralding the demise of "utility-based recreation." Finally, say writers like Paul Schullery, a former editor of *Yellowstone Science*, people are able to get an "educated look at free rising trout," and *of course* bans on fishing "induce a more respectful mood among visitors." Also, interpretive programs about the bridge and lake routinely emphasize "the inherent worth of and beauty of fish as wild animals rather than sporting quarry."

In 1994—the year the first mackinaw were discovered in Yellowstone Lake, when the cutthroat spawning run remained at its magnificent peak—about 170,000 people used the Fishing Bridge for "fish watching," spending an average of eleven minutes there. A similar number watched the spawning fish at LeHardy Rapids (a little farther downstream), and they hung around for an average of seven minutes.

Can a few minutes spent watching spawners really be enough to divine the inherent worth and beauty of fish?

In any case, all the angst seems to have amounted to nothing.

I should have braced myself for the worst. Instead I had ignored a recent article in *Yellowstone Science*, which suggested that "the waters are now essentially bereft of those fabulous fish," before going on to state that "this happened entirely because of the human agency of exotic species."

Are the catastrophes getting faster, or is it an illusion? Is humanity, and more importantly are my people—paleface, whitefella, pakeha—more open-minded or less open-minded than they were in Custer's time?

When you are ten years old in a secure Western society and you read about historical sadness—the sort of unbearable sadness that consumes whole communities, whole races—it feels as remote in time as the Cretaceous mass extinction or the Big Bang, as incomprehensible as catastrophe. Surely the causes of such sadness belong to a less enlightened time, to an archaic value system.

When I was ten, the Battle of the Little Bighorn had occurred almost ten of my lifetimes ago, and those lifetimes were inevitably measured at childhood pace. I didn't even stop to think, much less comprehend, that when I was born there were people still alive who had witnessed the battle.

Now I am fifty and the Battle of the Little Bighorn has occurred less than three of my current lifetimes ago, and those lifetimes are measured at adult pace. These days Custer no longer belongs to an incomprehensible past. And what brings Custer even closer to home is that *catastrophe* is no longer a mere concept to me.

At a personal level I've lived through disasters as big and tragic as *The Last Hunt* or the demise of the Bighorn cutthroats: the construction of a house on my childhood frog pond; the desecration of my grand Lake Sorell; the blooming of didymo throughout the South Island's once-sparkling streams; the massive dieback of pine trees in British Columbia's forests; the loss of glaciers in South America's incomparable lake country; the gradual but relentless mining of land and soul in Mongolia's vast, unfenced steppe.

Yellowstone just happens to be the latest catastrophe. I realize that the real reason I am here is that I have somehow become a sort of war correspondent—that I have developed a compulsion to bear witness.

Greg French

Chapter 7:
Yellowstone Lake

LAKE FISH
HATCHERY

Looking south over the lake from the Fishing Bridge, we see a big storm brewing. I want to see the historic hatcheries at Lake Village before we head into the backcountry again, and we'd better do it quickly lest we get struck by lightning.

Notwithstanding a checkered legacy worldwide, I confess to having an affinity for historical hatcheries. I spent several years as a hatchery officer at Tasmania's Salmon Ponds, the site of the first successful introduction of salmonids to the Southern Hemisphere. It remains the oldest trout hatchery in the Antipodes, and one of the oldest continually operated hatcheries in the world. I adore the weatherboard buildings, the quaint old Huon pine hatching troughs, the tranquil lawns dotted with exotic Northern Hemisphere trees, the thriving waterweeds in the earthen rearing ponds. To this day the Salmon Ponds has maintained the serenity of an English country estate.

I confess, too, that I empathize with the original intent that underpinned the construction and operation of the Salmon Ponds and Yellowstone hatcheries: a desire to help improve, expand, and diversify fishing opportunities.

The US Bureau of Fisheries established several hatcheries in Yellowstone National Park in the early 1900s, including one at Soda Butte Creek and the more famous one I seek at Yellowstone Lake.

The Lake Fish Hatchery Historic District is succinctly named: the fish hatchery was built on the shore of Yellowstone Lake. It is supposed to comprise nine buildings built from 1930 to 1932. Here, I have read, cutthroat eggs were collected and on-grown mainly for distribution outside of Yellowstone. Indeed, between 1901 and 1953 an estimated 800 million ova were exported to other hatcheries in the United States.

The National Park Service began winding back hatchery operations in 1953, ostensibly because of the negative impacts on native populations of westslope and Yellowstone cutthroats, but also because by this time anglers themselves were beginning to wonder if stocking was doing more harm than good. The last hatchery closed in 1957, and then jurisdiction for the buildings passed from the US Fish and Wildlife Service to the National Park Service.

In 1985, the Lake Fish Hatchery Historic District was included on the National Register of Historic Places. But you wouldn't know it. We've been driving around for half an hour now and have found no signs telling us where to go. It's as if local managers, embarrassed about what happened here, have deliberately tried to erase this part of Yellowstone's history. I feel … *betrayed*? Yes, that's the word.

Frances has just spotted a promising relic: a single-story building comprised of full-length round logs, notched at the corners and chinked with mortar. We pull up alongside the main entrance. Weeds grow tall against the foundations. The timber is so old that it looks polished. We get out of the car and peer through the windows. The room is completely empty and spotlessly clean. Log rafters radiate from a great central stone fireplace. Behind a large built-in reception desk, a hallway leads to a battery of office doorways. A park ranger walks out of the hallway past the desk, and we accost him at the main door. No, he says, this is not one of the hatchery buildings; rather, it's the original visitor center, recently retired from service. And no, he doesn't know where the hatchery buildings are. But he wishes us a nice day.

We get back into the car. Surely someone at the Lake Hotel will know where the hatchery district is.

Greg French

Built in 1891, the huge hotel is the oldest building in the park. Here the designers eschewed raw logs in favor of civilized clapboard and affected grandeur. Ah, the crisp modernity of the cladding, painted pastel yellow, and the regimented neatness of the small windows, trimmed in white. Someone also saw a need to include two large porticos supported by Greek columns.

The Lake Hotel reminds me of a plantation owner's mansion: a statement of authority, and dominance. But somehow the building's maturity has taken the edge off the designers' intentions. Instead of garish pretentiousness, the building radiates a type of beautiful melancholy, a mood enhanced by the unkempt grounds and the moss that infests the old timber roof shingles.

We drive closer to the hotel, past a series of pastel-yellow cabins. They seem neglected in their ramshackle overgrown meadow, though on closer inspection all are occupied by happy travelers. Finally we reach the once-grand circular driveway fronting the hotel's entrance foyer. Spindly weeds grow from cracks in the crumbling bitumen and concrete. We get out of the car and look up at the roof with its multiple gables and tiny dormers. The gutters are rusted and leaky, the paintwork blistered and peeling. Such a contrast to all the neatness at Mammoth.

We make our way up three wide concrete steps, across a concrete pavement, into an entirely different world. The interior is meticulously maintained, the staff neatly uniformed and grandly professional. I cross the sparkling floor to the reception desk and ask about the hatchery.

"I am terribly sorry, but I don't think I can help you. No one has ever asked about the hatchery before. It's on the lakeshore somewhere, I think. It might be best to inquire at the visitor center. Hope you find what you are looking for. Have a nice day, won't you?"

Eventually we do find what we are looking for, though the only signage says, "It is an offense to enter locked buildings or damage government-owned property." I am able to confirm the identity of the buildings only by comparing what I see to photos and a description from Wikipedia, which I printed out before leaving home. Yes, "the buildings exhibit a consistency of style and construction, with exposed gable trusses and oversized paired logs at the corners, all with brown paint." And yes, "the chief building is built with a 'logs out' technique of construction, in which the log frame is exposed on the outside and the sheathing set in." But does the interior have the requisite "smooth finish"? I peer through gaps in the shuttered windows. Yes, the interior does indeed have a smooth timber finish, but other than that, there's just empty space.

There is still some afternoon left, and despite the weather I want to fish.

We drive back through Lake Village, past the Bridge Bay campground—where we have reserved a nook for the night—and try to find a protected piece of lakeshore.

MAN, GIRL, STORM

Wind buffets the car and rain flushes in waves across the windscreen. We slow down and end up parking in a vacant pullout along the road between Gull Point and Sand Point. Mathews and Molinero say this stretch of shore is one of their favorite places to fish Yellowstone Lake, and favorite or not, today it is the only place that might offer any semblance of shelter. Frances looks at me askance and says she'll be happy enough reading her novel. Normally on an afternoon like this we would retreat to a motel and remind ourselves of why we have stayed married for more than two decades. But Yellowstone Lake is the epicenter of the cutthroat problem, and essential to the story I'm trying to write. Our itinerary is tight enough as it is. If I don't fish on the lake now, I might not get another chance.

The beach is relatively sheltered, but the rain remains torrential. I can hear nothing above the roar of wind, certainly not the crunch of the gravel beneath my feet, though I know that soon enough I

123

will hear more trees crashing down in the pine forests behind me, more hideous thunder over the water in front of me. I look eastward toward the dark gray horizon. Any delineation between lake and sky has been obliterated.

I hold the fly rod vertically in the wind. The fly is whipped from the keep, up into the air out toward the water. It tugs at the leader. I release some fly line from the reel, and it feeds itself up through the guides out into the gale. Soon five yards of line are fluttering horizontally in the air. I lower the rod tip, and the heavy Black Woolly Bugger alights fairy-soft on the water.

A low cliff behind me creates a small windbreak, and a six-foot-wide band of calmness extends along a couple of hundred yards of beach. Although the air is as dark as it can possibly be at two in the afternoon, I can see quite well along the edge. There are no fish. I cast beyond the calm water onto the seam where the wavelets begin. After all, seams define currents, and currents concentrate food. If nothing happens, I'll try casting ten yards farther out where the whitecaps begin.

I never enjoy fishing blind and have no patience for standing still, so I move northward along the shore in the direction of the best visibility in the forlorn hope I might yet spot something.

Ahead of me, sitting on a driftwood log, in the middle of the gray gravel, lapped by gray water, in gray air, is a person. He's about fifty yards away, which is as far as can be seen in the seething, wet air. I fish toward him but he doesn't notice me.

When I'm five yards from him, I shout, "Hello!" but he doesn't move. Perhaps the wind has carried my voice offshore.

I skirt around him, cautiously, as if he were a bear.

A fly rod rests on his lap, and I can see that he is staring at something. A wallet? He has the hood of his gray rain jacket pulled over his head, and I can hardly see anything of his face, yet I am almost certain that he is crying. If so, it's a stoic sob, one that doesn't heave his shoulders or even flutter his chest.

He is *unnaturally* still in the storm—not even the buffeting wind moves him. I wonder how I know he is sad. Does grief have a smell? My eyes prickle.

I wander up the beach, away from confrontation, periodically looking back over my shoulder, until the sadness is merely a smudge in the gloom, and then I go back down to the lakeshore. I am distracted by what I have seen, and I don't recall casting, but suddenly, miraculously, weight is tugging on the end of the line, followed by boiling water and line peeling from the reel.

By the time the fish is beached, the wind has died away and the rain has become a fine mist, almost fog. It's a beautiful fish, chunky and strong, about four pounds. A sunbeam of bright yellow amid the depressing gray. I wish I had brought my camera, but it's hard to take professional shots when fishing alone—and even harder in a storm.

"A good one," says a voice at my side.

I turn to see the man, rod in one hand, wallet in the other. The wallet is still open and I can see a portrait photo behind a plastic window. It's a girl, maybe fifteen, maybe younger. She has long golden hair and a sprinkling of large freckles. The freckles look too dark. Perhaps they are not freckles, merely a trick of the light upon raindrops or tears.

"That's Claire you're looking at," he says with the slightest hint of offense, and I am embarrassed.

I look up into his eyes. The hood is off his head. He is older than I expect: my age or a bit older. Fifty-five? His hair is darker than mine, but he is balding on top. Slightly overweight but fit. We are not so different from one another.

"She's beautiful," I say, aware that whatever I say will intrude upon his grief. "Your daughter?" I sound clumsy.

There is a pause.

"My lover," he says.

I think of Lolita. Was her hair yellow? I turn my head away. There is an ugly silence.

"You should let that fish go."

I do. We watch it swim off.

"Precious," he says. I assume that he's talking about the fish, but when I look at him, he is studying the picture in his wallet.

"You fish here much?" I ask, because I must say something.

"I like the rivers best: Buffalo Ford, the Lamar, Slough Creek. Claire always liked the lake, spots like this." He points a finger toward the place in the lake where we both suspect my fish should be by now. "She never caught one that big."

Claire in the past tense. I try to think of something to say, and the silence becomes deafening in my ears.

"We don't catch so many cutthroats as we used to," he says. "But the fish are bigger than they were, more precious than I ever guessed."

Claire in the present tense? Is she alive or not? The rain has completely stopped, so the wetness on his cheeks must be tears. He tries talking to me as if he were not crying. "I suppose it pays to think about what we have, not what we've lost, what we can still do, not what we shouldn't have done or what we could have done instead." He sounds distant, as if the effort of talking to me has become too much, as if it's easier or more comforting to have a conversation with himself.

Lightning. A deafening clap of thunder. Ozone. The rain starts to pour again, and I say, "Well, I best be going."

He holds out his hand, and I shake it. "Have a nice day," he says.

"My wife's in the car," I say apologetically, and walk off down the beach.

BRIDGE BAY

We drive back toward Bridge Bay, and experience several long delays while road crews clear fallen pines. I give a blow-by-blow account of what happened with the man on the lakeshore.

"Don't make him sound so sinister," Frances says. "It sounds to me as though the photo might be one of his wife when she was young. He's probably carried it around in his wallet ever since they first met."

We're nearly back at Bridge Bay now, and I'm dreading using a major road-accessible campground. At home this would mean having to endure the constant thrum of generators, the feigned outrage of shock jocks emanating from unrestrained radios, camper vans blocking out views and sunlight, and many more loud and garish affronts. I wish we had been able to get a room or cabin when we enquired more than six months ago.

The rain has stopped, and the wind seems to be dying. We pull up alongside an entry booth and pass our reservation printout to the attendant. She asks us how we have been enjoying the park. We talk amicably about hiking and wildlife, architecture and fishing. Then she gives us a map of the camping complex, inks in the most direct route to our campsite, and tells us that we can buy a small bundle of kindling and a box of firewood from the adjacent kiosk. The shop is reasonably well provisioned, and we buy some fresh milk as well.

We drive past a big, flat meadow set aside for Winnebagos and trailer tents. Next we pass a lightly wooded area set aside for large four-wheel drive vehicles and family tents. Now we are going uphill through a dense forest of Engelmann spruce. The road remains surfaced with bitumen but has become narrow and strictly one-way. There are numerous cozy nooks, and I am suddenly grateful that when making our reservation we stipulated that we'd arrive in a small car and sleep in a hikers' tent.

Despite being just a few hundred yards from the nearest amenities block, our alcove feels more like a nest than a campsite. The nearest campers are thirty yards away, and apart from the welcome giggle of young children, we really don't notice them. The only other sounds come from busy squirrels and chipmunks.

Greg French

PELICAN CREEK

By early morning the weather has cleared. Later today we will walk to Heart Lake, but what should we do in the meantime?

My original intention had been to hike up Pelican Creek, which enters the lake less than two miles east of the Fishing Bridge.

Mathews and Molinero note that Pelican Creek is closed to fishing for the first two miles upstream of the lake to protect spawning cutthroat trout, but recommend hiking along the Pelican Creek Trail because good numbers of spawning fish can be expected in the mid-reaches from opening day (July 15) to late August, by which time most fish will have dropped back to the lake.

The problem is, their book was published in 1997. When I borrowed the waitress's laptop in Gardiner—after the bloke at the Parks' Fly Shop suggested that Pelican Creek might be closed to angling—I looked at several National Park Service articles. In the mid-1990s, apparently, the spawning run in Pelican Creek comprised around 55,000 fish, but over the last decade the number of spawners has fallen to around 500. The situation is now so bad that there is talk of a "trophic cascade," whereby animals further up the food chain—everything from pelicans and ospreys to otters and bears—are having a pretty hard time surviving.

Why have the fish vanished? The quantity of spawners in most other tributaries has not dropped quite so dramatically, yet fish born in those tributaries are not migrating into Pelican Creek to take up the shortfall. According to the National Park Service, the main problem seems to be whirling disease, which was first detected in trout from Yellowstone Lake in 1998.

Halverson gives a pretty good history of whirling disease in *An Entirely Synthetic Fish.* It is caused by a tiny parasite, *Myxobolus cerebralis,* native to Europe. European salmonids, including Atlantic salmon and brown trout, co-evolved with the parasite and are highly resistant to it. American salmonids, particularly rainbow trout and cutthroat trout, are highly susceptible to it.

The parasite latches onto newly hatched fry, bores through the skin, and ejects germ cells (reproduction cells) that burrow into the flesh looking for cartilage, primarily in the skull and spinal cord. Young fish are especially vulnerable because the cartilage in their skeletons has not yet been replaced by bone. The spores do all sorts of neurological damage, and often cause stroke-like paralysis down one side of the body, which, predictably, causes the fish to swim in circles. When the fish die, the spores are consumed by *Tubifex* worms. Within the worm, the parasite grows to maturity, ready to find more fish.

According to Halverson, the disease was first noticed in Colorado hatcheries in the winter of 1987. There was an initial quarantine period, but the disease had been evident in other American states since the 1950s and did not seem to present any problems in the wild, so by summer it was back to business as usual, with Colorado hatcheries stocking fish that were known to be infected into public waters, even waters that supported robust populations of wild trout.

In autumn 1993 it was discovered that there were hardly any two- and three-year-old rainbow trout in the upper reaches of the Colorado River. This turned out to be the first outbreak of whirling disease in the wild, one that resulted in a 99 percent mortality of rainbow trout.

Colorado's response to the problem was to look for disease-resistant stock, which it eventually located at a German hatchery operated by the Hofer family. The problem is that the Hofer rainbows are domestic animals from a very limited gene pool, which almost certainly leaves them and their hybrid offspring vulnerable to outbreaks of other diseases and less able to adapt to new environmental challenges such as climate change.

Montana immediately began testing for whirling disease in its own wild trout fisheries, and sure enough, they found it. How it got to Montana no one knows, though some say it may have been transported by fish-eating birds or even on someone's wading boots.

Greg French

The Madison was hit especially hard and the rainbows disappeared virtually overnight, resulting in the river's third hatchery-induced disaster. Nonetheless, Montana refused to supplement the dwindling stocks of wild fish with hatchery replacements, and by 2007 strong evidence showed that what was left of the wild stock had developed some natural resistance. It is true that most fish still die early, usually as they reach sexual maturity, and that the future is uncertain, but there is plenty of hope.

On the Internet in Gardiner I could not find any agreed-upon reason why whirling disease should be more prevalent in Pelican Creek than other tributaries of Yellowstone Lake. So I used Google's satellite imagery to home in on the stream mouth. I immediately realized that spawning cutthroats from the lake would have to migrate over a vast, silty lagoon complex, which looked like ideal habitat for *Tubifex* worms. Remembering Chuck's comments at Buffalo Ford, I wonder if the silt problem at the Pelican Creek river mouth has been exacerbated by erosion following the catastrophic 1988 fires—the silt load from the first and subsequent floods must have been enormous. Judging by the persistence of colossal sandbars at the mouth of Tasmania's King and Ringarooma Rivers more than a century after mining operations ceased, I doubt that Pelican Creek will return to normal any time soon.

Back in Gardiner, I looked up Mathews and Molinero for another lake tributary, and Clear Creek sounded good. But the Internet said that the number of spawning cutthroats migrating up this water had plummeted from more than 70,000 in 1978 to less than 7,000 in 2002 to around 500 annually since 2006. The Google Earth image of the stream mouth suggested no big silt complex this time.

In any case, there doesn't seem to be much sense in fishing the lake tributaries. We'll have to think of somewhere else to go. Maybe the lake itself will be worth a try.

YELLOWSTONE LAKE

The clouds are high and thin and Yellowstone Lake is vast and beautiful, so we decide to fish Bridge Bay.

It's hard to imagine the lake in early December when it is freshly frozen over, and harder still in mid-winter when the ice is three feet thick and the only free water is located in shallow areas atop hot springs, the way it was when John Colter arrived in 1807–08.

Colter is said to be the first person of European ancestry to have explored what is now Yellowstone National Park. He was a member of the Lewis and Clark expedition (1804–1806), and afterward he decided to try his hand at trapping beavers, which is how he ended up here.

More than two centuries have passed since Colter arrived, and considering what Europeans have done elsewhere in America, I'm surprised that the lake remains in its natural state. It certainly wasn't for lack of ideas about how to modify it, or for a lack of justification either. Between 1920 and 1937 there were many proposals to dam the outlet so that floods could be controlled, so that water could be diverted into the Snake, so that all manner of "progressive" ideas could happen. I find it incredible that none of these proposals ever got through Congress, but perhaps that's because I grew up in Tasmania at a time when our state-run Hydro Electric Commission was quite unstoppable in its efforts to inundate national parks and other areas of high conservation value. I find it unbearably sad that, despite Americans having taken decisive action to preserve the upper Yellowstone catchment, the aquatic ecology is being destroyed by default.

Will I target cutthroats or mackinaw?

The first time I saw a mackinaw was in the early 1990s at Lake Pearson, beside the main road between Christchurch and Arthur's Pass National Park on New Zealand's South Island. It was late autumn and I had been flats-style polaroiding for cruising brown trout at the head of the northern basin when all of a sudden the only other person in sight, a spin fisherman, approached me and asked if I knew what sort of trout he had just caught.

I had recently bought and read R. M. (Bob) McDowall's authoritative *New Zealand Freshwater Fishes* (published in 1990), and immediately realized I was looking at a mackinaw. It wasn't a great specimen. It was old and very dark—so much so that you could barely make out its pale yellow spots—with a big head and somewhat slabby body; a female that looked as if it had just spawned. We both guessed that it weighed little more than a pound.

McDowall said that the species had been introduced to New Zealand from Michigan in 1906. Liberations were promptly made in Lake Pearson and nearby Lake Grassmere, then promptly forgotten. The Grassmere population died out in the 1970s, but the species hangs on at Pearson, the only wild population in Australasia. A dozen or so fish are caught each year, usually in late autumn or spring. And the reason the fish are small is that Pearson is small. According to McDowall, "Mackinaw require very cold water and very high levels of dissolved oxygen, and are normally associated with large, deep lakes." Like Yellowstone.

Last night I reread Behnke, and gleaned a wealth of interesting trivia about what the species *should* be like. As mackinaw mature, most become highly piscivorous. The oldest known specimen was sixty-two when caught, the next oldest fifty-three, but in many waters most fish are aged ten years or less. The largest "laker" ever recorded weighed 102 pounds—and there are many records of fish over sixty pounds—but most weigh less than ten.

Since mackinaw sit at the very top of the food chain in nutrient-poor lakes, they exist in low numbers and are vulnerable to overexploitation. The most prolific mackinaw lakes support just two to four pounds of fish per surface acre, whereas good brown or rainbow trout waters support up to 500 pounds per acre.

There was a famous commercial fishery for mackinaw in the upper Great Lakes (where they are native) in the early 1900s, and even though the annual harvest never amounted to more than half

a pound per acre, that was enough to send the population into dramatic decline.

If mackinaw populations are so vulnerable, why, I wonder, isn't the suppression program working in Yellowstone Lake?

In *About Trout*, Behnke contradicted the idea that the lake's native cutthroats and nonnative mackinaw always utilize separate microhabitats. Apparently young cutthroats of less than twelve inches are rarely found in Yellowstone Lake, and Behnke speculated that once the tiny hatchlings have migrated from their redds in the spawning rivers, they probably spend time deep down in the open water well offshore. He also noted that fingerling cutthroat trout are not found in the guts of adult cutthroat trout (even though big cutthroats are cannibalistic, being famous for gorging on accumulations of fry at the river mouths), and that cutthroat trout do not move into the shoreline shallows until the third or fourth year of life.

Because Yellowstone Lake is nutrient-poor and often iced over, there is a scarcity of food. Poor nourishment affects egg production, and consequently mature cutthroats only spawn every second year. Despite the fact that Yellowstone Lake's cutthroat trout eat plenty of insects, especially in summer, the bulk of their diet comprises crustaceans, primarily planktonic daphnia (water fleas), and bottom-dwelling amphipods (scud).

(The relationship between feeding seasons and trout fertility explains why trout have failed to establish viable populations in many alpine waters throughout North America. Usually it's not because the lakes freeze to the bottom or that waterweeds suck oxygen out of the water during ice-over, as is commonly assumed, but because the fish never mature.)

Now, ever since I saw my first mackinaw at Lake Pearson I have wanted to catch one on a fly. Before leaving Tasmania, I even considered hiring a local Yellowstone outfitter to take me out on the lake. But since the mackinaw live deep down and I would have no opportunity for stalking my prey, I quickly lost interest. Why blind-fish for introduced mackinaw when I could be polaroiding natives?

I wonder why catching native fish in native environments is so important to me. After all, brown trout aren't native to Tasmania, and I still consider the fishing at home to be equal to the best in the world. Perhaps I like Tasmanian browns because we have no native sportfish. Perhaps it's because these days our brown trout are more genetically diverse and wilder that most brown trout in Europe. If that sounds like rationality being trumped by fervor, I suspect you are right but, hey, we're all filled with contradictions. Take Dale, a vegan friend of mine who won't eat honey because he doesn't like upsetting insects. I once tramped with him up a backcountry stream on the South Island of New Zealand, and I swear he swatted a million sand flies.

Currently Dale is in Yellowstone Park working out ways to kill nonnative trout. Before leaving home, I wanted to know if the stomach contents of harvested mackinaw were being systematically examined by scientists working on the mackinaw-suppression program. More specifically, I wanted to know whether anyone knew if the consumption of cutthroat trout by mackinaw was reducing over time, and if so, what alternative food might be filling the void. Dale directed me to several Yellowstone scientists, field assistants, and bureaucrats, but unfortunately I didn't get any response to the e-mails I sent them.

Dale also e-mailed me the proof that mackinaw were translocated to Yellowstone Lake from Lewis Lake in 1989: a paper titled *Natural chemical markers identify source and date of introduction of an exotic species: lake trout* (Salvelinus namaycush) *in Yellowstone Lake*, written by Andrew R. Munro et al. in 2005.

The authors recorded the strontium-to-calcium ratios in the ear bones of mackinaw collected from Yellowstone Lake during the early stages of the mackinaw-removal program in 1996 and 1997. They then compared these ratios to those of mackinaw collected from Lewis Lake and Heart Lake, which they assumed would be the source of any translocated stock.

The twenty fish that they sampled had been randomly selected from 164 of the biggest, and were aged from thirteen to thirty-two years. Eighteen were found to exhibit a sudden change in relative abundance of strontium and calcium, consistent with being transplanted from Lewis Lake. Only three fish of the eighteen were tested to assess the exact timing of the change. Of these, two were estimated to have been transplanted in 1989 and one in 1996.

What I found most intriguing about the paper were the questions that hadn't been asked. Foremost: since the idea of at least two mass translocations of adult mackinaw from Lewis Lake is so implausible, why hasn't anyone tried looking for other ways to explain the data?

Could it be that there was a decades-old sleeper population in the lake? After all, prior to 1988 Yellowstone Lake was a low-nutrient lake, so the mackinaw population would necessarily have been small. And given the species' propensity to live at depth, it could only be noticed if someone thought to look for it, and why would they do that if they didn't have any inkling it was there?

Could the massive increase in mackinaw numbers have been precipitated by a subtle increase in nutrient levels resulting from inflowing silt and ash?

One way to test the "transplant theory" would be to compare the strontium-to-calcium ratios in the ear bones of the supposed transplants to those of cutthroat trout living in the lake prior to 1988. All you would need to do is locate some archival specimens of cutthroat trout—perhaps fish that have been preserved during previous studies, or mounted specimens in anglers' private collections.

No one seems to have remarked on the fact that, although mackinaw normally exist in relatively low numbers, the population in Yellowstone Lake is high and getting higher. Sure, a population boom is common when nonnative salmonids are first introduced into a lake, but after more than two decades, and the substantial loss of the initial food resource (cutthroat trout), why haven't mackinaw numbers stabilized or even crashed? After all, mackinaw are typically overexploited even when fishing pressure is very low, and the local authorities at Yellowstone claim to be removing hundreds of thousands of the damned things each year.

Could it be that the decline of the cutthroat has mostly been caused by whirling disease, and the mackinaw population has simply expanded to fill the void?

Perhaps the trout have been affected by changes in the invertebrate population. Cutthroat trout feed heavily on daphnia, and daphnia feed on phytoplankton (planktonic plants). It has been documented that as the cutthroat population nosedived, the daphnia population skyrocketed, resulting in a greater harvest of phytoplankton, apparently enough to increase the clarity of the lake by five feet or more. Could the lake now be more highly oxygenated than it was prior to 1988? Could this have made the lake more suited to mackinaw than it was before 1988? Could some other feedback loop have been triggered?

David Sweet, of the East Yellowstone chapter of Trout Unlimited, told me that Yellowstone mackinaw feed heavily on bottom-dwelling scud. This must be why mackinaw are currently able to outnumber cutthroats, but how long can a top-order predator maintain a higher biomass than a second-level predator like cutthroat trout?

The mackinaw-suppression program bothers me because it is not suppressing the mackinaw. I'm not entirely surprised. Remember, in Tasmania, trout stocks are usually limited by the carrying capacity of the lakes rather than recruitment; the more fish removed from the water, the greater the survival of naturally spawned fingerlings. The huge annual harvest of brown trout by recreational anglers in our most-fished waters has no observable or measurable impact on fish stocks.

The population of mackinaw in Yellowstone Lake has been increasing exponentially and must now be approaching capacity. Surely much more biological energy is required to grow a replacement crop of top-order predators than to sustain fish that have already grown to optimum size?

These are the questions I want to discuss with Behnke when I interview him in two weeks, because given the timing and the nature of the mackinaw problem, I still think a link between the 1988 fires and the "ecological collapse" in Yellowstone Lake is highly probable. The National Park Service website insists that wildfires occurred across much of the Yellowstone ecosystem in the 1700s and are perfectly natural. The problem, however, is that mackinaw and whirling disease did not exist here back then. And the park itself is now an island of wilderness, no longer part of an almost-infinite whole. Using the 1700s as a benchmark of what *natural* should be ignores the reality that our planet has fundamentally changed. If the preservation of biodiversity truly is the main aim, I wonder if allowing the "natural" processes of the 1700s to hold sway in the 2000s could turn out to be naïve stewardship.

We conservationists are preservationists at heart, and we find it much more comforting to postulate that "irresponsible rednecks" have undertaken an illegal liberation of mackinaw than to acknowledge the true complexity of managing wild landscapes in modern technological times when the introduction of invasive species, the fragmentation of wilderness, and the advent of rapid climate change have forever altered old paradigms.

Bloody mackinaw. The last thing I want to do right now is go fishing for them. Anyway, my booklet of Yellowstone angling regulations insists that if I did catch a mackinaw, I would be legally required to kill it, preferably by puncturing its air bladder and letting it go, and such nastiness is an anathema to me.

From our vantage beside the road paralleling the northwestern shore of Bridge Bay, we stare down into the water. A southerly wind has sprung up. The water is choppy and the air has filled with smoke from yesterday's lightning strikes. Polaroiding is going to be next to impossible.

I look offshore and study a few promising froth lines. Hey, I can discern the odd good rise. I gaze out farther, off the northern tip of Stevenson Island, and see that a fantastic wind lane is forming. It will be concentrating food as well as froth, and by rights, fish should be going crazy out there.

Mathews and Molinero say there is no need for a boat on Yellowstone Lake, but I suspect this is only because hardly anyone in Montana and Wyoming fishes lakes the way we do in Tasmania—the stream fishing is so good that it has become culturally dominant. I have another thought too: Montana and Wyoming are landlocked, and the locals don't get to play in the surf during their formative years, don't get to feel events like longshore drift and undertows, and as a result, they probably don't suspect how dynamic lakes are.

All of a sudden I have a compulsion to get out to Stevenson Island. I wonder if the cutthroats in the wind lanes behave like rainbows back home, filter feeding in the manner of basking sharks, swimming open-mouthed through dense accumulations of daphnia. Perhaps this is where I will find juvenile cutthroats too, not down deep but right on the surface.

LAKE CURRENTS

In my home state of Tasmania, the best trout fishing, or at least the quintessential fly fishing, is lake-based. We call lakes "still waters," but like restless hearts, they are laced with currents, rips, undertows, and upwellings.

In his book *The Wayfinders: Why Ancient Wisdom Matters in the Modern World*, Canadian anthropologist Wade Davis described a trip he took aboard the Polynesian boat *Hōkūleʻa*. The Polynesian sailors, he discovered, could sense at least five different types of sea swell. By putting an ear to the hull, a man could hear the deep, persistent ocean currents, and he could follow them "as easily as a terrestrial explorer might follow a river." They could also sense distant islands by virtue of the echoing surface currents, and identify each island so heard because different landmasses leave different oceanic fingerprints in the waves.

Davis stressed that different lifestyles, passions, and cultural imperatives enable ways of thinking and perceiving that are not available, or even possible, to those confined to urban environments and burdened by modern prejudices. He said that to argue for the cessation of religious, cultural, or tribal practices (and dare I say fly fishing) is to desecrate humanity's achievements and deplete humanity's wellsprings of understanding.

I first realized how different passions and cultural imperatives enable different ways of thinking and perceiving when a team of English fly fishers came to Tasmania for the 1988 World Fly Fishing Championships and proceeded to demonstrate to local writers and fishing guides what the English could achieve on the lakes "Down Under."

In the same way that English fly fishers were able to see things differently than Tasmanian fly fishers, I soon found that Tasmanian fly fishers were able to see things differently than New Zealand fly fishers. In 1997, I bought an issue of *Fish&Game* magazine, which included an article summarizing a recent study of Lake Coleridge's wild salmonid populations: researchers had radio-tagged a dozen fish of each species (brown, rainbow, quinnat) and mapped their movements. Apparently the brown trout cruised along at less than 0.1 miles per hour and rarely moved more than a couple of hundred yards from home, which was something every experienced polaroider would have expected. The rainbow trout stayed well offshore and cruised in relatively straight lines or arcs at more than 0.2 miles per hour, often covering "considerable distances," which was something every experienced wind lane exponent would have expected. The quinnat salmon—and this is what really piqued my interest—were also swimming in relatively straight lines or arcs, but at an average speed of more than 0.5 miles per hour, and had no trouble covering the eleven-mile length of the lake in a single day.

The stomachs of the quinnat salmon contained mostly terrestrial insects, yet the researchers concluded that the "roving" behavior of the quinnat salmon must be associated with the fish's search for the larvae of pelagic bully (a type of small fish), since bully were known to occur in moderate numbers in Lake Coleridge between the surface and mid-water over the summer months.

I asked a lot of experienced Tasmanian fly fishers to interpret the data, and every single one of them immediately drew a conclusion quite different from that of the New Zealand researchers: surely the quinnat salmon were taking beetles from the surface in wind lanes just like typical rainbows, only moving much faster and probably feeding much more efficiently.

Culture and experience really do affect what a researcher is *able* to comprehend. And I am certain that my Tasmanian fly-fishing background will enable me to take advantage of a bonanza in Yellowstone Lake which few, if any, locals have properly exploited. Perhaps I really will discover the elusive feeding grounds of those never-seen juvenile cutthroats.

Frances and I turn around and go back to the Bridge Bay marina.

"I'm very sorry, but all our boats are taken. If you want to come back in a couple of hours…"

Ah well, it was worth a try. Anyway, it's probably best that we head off to Heart Lake sooner rather than later because this will allow extra time to examine another part of the puzzle.

Chapter 8: The Snake River Catchment

LEWIS LAKE

The road to Heart Lake follows the western shore of Yellowstone Lake's West Thumb. Almost as soon as you leave the southwestern corner of the bay, you cross the Continental Divide into the Snake River catchment. Soon after that you hit the eastern shore of Lewis Lake.

Lewis Lake (along with interconnected Shoshone Lake), on the Lewis River upstream of Lewis Falls, originally contained no fish whatsoever. According to my guidebooks and the park website, the first-ever releases of hatchery-reared fish in Yellowstone Park occurred here in 1890, and included some 42,000 mackinaw fingerlings, which soon established a vibrant breeding population.

These days, in summer at least, most mackinaw are targeted with leech patterns and sinking lines along the drop-off flanking the southwest shore, but Mathews and Molinero say you can sometimes hook one or two on a dry fly. Cool.

Also released into Lewis Lake and Shoshone Lake in 1890 were European brown trout. Behnke says that these were progeny of fish sourced from Scotland's Loch Leven, and that there are no further records of any waters in the upper Lewis system being stocked with brown trout. This interests me even more than the mackinaw because there are very few genetically pure Loch Leven fish left anywhere in the world, perhaps not even in Loch Leven itself.

The first time I considered going to Loch Leven was in 2005, but I was disappointed to learn that the water had become thick with algae and was being managed as a hatchery-reliant rainbow fishery.

When, inevitably, the rainbow experiment failed, the focus shifted to environmental rehabilitation, and that brought about a resurgence of naturally spawned brown trout. So I recently went back there and caught a couple of handsome silvery fish. Unfortunately, because the lake's hatchery had been in and out of operation for well over a hundred years, none of the experts I spoke to could guarantee the genetic integrity of the fish I caught.

The original Loch Leven stock were distinguished by being rather elongate and having a relatively silver body heavily marked with black spots and no red spots. They also had large fins. Historically they averaged about one pound, though fish to two pounds were common and some grew much bigger. In the mid- to late-1800s they were considered to be a unique species, *Salmo levenensis*, and their superior sporting qualities meant that they were in demand worldwide. By the late 1800s live ova had been shipped to many countries and colonies, including New Zealand, America, and Tasmania.

In *Trout Fishing*, Joe Brooks said that some anglers in the United States still thought they could distinguish Loch Leven trout from the "normal" German brown trout. And in Tasmania when I was growing up, all the old hands at Lake Sorell also thought they could distinguish Loch Leven trout from other varieties. (It is most likely that these people were wrong, that the fish they thought were different were actually hybrids.)

Today the Loch Leven trout is classified as *Salmo trutta*, the same as all other brown trout in New Zealand, America, and Tasmania. But, same species or not, there is no denying that pure-blooded Loch Leven trout were distinctive and culturally important. And perhaps Lewis Lake is as close as I will ever get to sampling the real thing.

Mathews and Molinero say that from about mid-June, the brown trout in Lewis Lake begin rising to midges and *Callibaetis* mayflies. The weather has warmed up a lot since mid-June, much more than is normal, and the fish may well have moved offshore into deeper, cooler water, but I strongly suspect that I could spot one or two

Greg French

cruising along the steeper shores and entice them to eat a dry or nymph. Frances, however, says that from the moment we decided to leave Yellowstone Lake as early as we did, she set her heart on getting to Heart Lake by early afternoon. "I'm only going to stop driving if we see fish rising steadily. You can't experience everything Yellowstone's got to offer when we are only here for a few weeks."

There are no rises.

HEART LAKE

Heart Lake, like Lewis Lake, lies in the upper Snake River catchment, but it isn't located above a barrier fall and therefore it supports a native population of Yellowstone cutthroat trout.

As far as is known, Heart Lake has never been deliberately stocked with nonnative fish, but mackinaw exist there nonetheless. It is widely assumed that soon after the 1890 liberation of mackinaw into Lewis Lake, some fish migrated down the Lewis River, over the falls, and then swam up the Snake and Heart Rivers.

Mackinaw are not commonly seen in rivers, but apparently a significant number of individuals migrate through fastwater environments because in geological time the only barriers to the expansion of the species' native range have been saltwater (to which mackinaw are singularly intolerant), warm water, barrier falls and, of course, land barriers.

Fishing guides advertise Heart Lake as an ideal place to target mackinaw on flies. Wilderness Trails Inc., for example, will take you on horseback to one of their designated campsites, and row you out over the deep water on a three-person raft. Unfortunately, the fishing entails sinking lines and weighted streamers. Frances and I want to stalk natives, and we want to do it alone. Unlike the situation in Yellowstone Lake, the native cutthroats in Heart Lake have happily coexisted with the mackinaw for maybe a century, and I'm keen to see firsthand how they've been able to do it.

The first few miles of the trail proves to be fairly flat and well shaded by lodgepole pines. Still, it's a hot blue-sky day, there's no water in the creeks, and a quart each of bottled water isn't enough. By the time we reach the halfway mark—the point where we get our first glimpse of the lake—we are desperate for a coldwater stream; instead we find a bubbling cauldron of sulfurous hot springs.

At least it's all downhill from here: a gentle two-mile descent to the northern beaches of the lake, then another mile or so along the western shore to our campsite.

Yes! Less than half a mile from the cauldron, the track crosses a magnificent mountain creek with grassy undercut banks and water that's crystal clear and invitingly cold.

No! The water is not at all cold. Mercifully it's not quite scalding either, otherwise I'd be bandaging my hands by now. I look on the map. The creek is sourced directly from the cauldron. Witch Creek.

It's a few hundred yards farther on before we find drinkable water in a tributary of Witch Creek; tepid, but good enough to quench our thirst.

I hate sterilizing water and usually opt to take my chances with *giardia* and *cryptosporidium*. Face it—iodine tablets taste revolting, and under current circumstances boiling the water would entirely defeat our desire. Anyway, in my travels around the world I've drunk water from streams far more suspicious than any I've seen here in Yellowstone. Nonetheless, Frances has invested in a battery-operated magic wand—a gadget that emits DNA-destroying ultraviolet light and prevents bugs from replicating in your gut—and ninety seconds later my cup of water is deemed fit to drink.

The lake is invitingly close. It's invitingly calm too, and I'm impatient to get a fly on the water. My guidebook says that summertime cutthroat fishing is good, but echoing in my ears is the advice from the Parks' Fly Shop in Gardiner: "The problem with our lakes is that they are too hot in summer and the fish retreat to the depths." Terrible seeds of doubt.

Greg French

We arrive at the lake's northern beaches. A thunderstorm is brewing, wavelets are turning into whitecaps, and the sky has become hazy with smoke from yesterday's lightning strikes. Since polaroiding has become next to impossible, I try prospecting with nymphs, but nothing happens. Well, something happens: lightning strikes a tree a few hundred yards from where I'm wading.

Frances and I seek shelter in a dense stand of pines. As the storm passes, a few tiny fish begin blipping offshore, but again I'm haunted by the advice from the Parks' Fly Shop: "If you do see rises at Heart Lake, most likely they'll only be Utah chub." When I asked, the shop assistant informed me that Utah chub are a cyprinid native to Heart Lake: *Gila atraria*. Apparently specimens of six to eight inches are typical, and the very biggest weigh less than two pounds.

Rather than continue fishing, we decide to make for camp. The consolation prize is the wildlife: garter snakes and giant frogs on the lakeshore, a bald eagle in a dead tree, a mule deer on the trail.

The camp is set in another copse of pines and well sheltered from the wind. A collection of log seats is arranged around a substantial fire pit, along with the obligatory bear pole. We cook a vegetarian stir-fry for dinner and then try our hand with weighted streamers along the drop-off and at the mouth of a nearby tributary creek. Nothing happens. Not even Muddlers at dusk do any good.

The moon comes up, and back at camp—while drinking coffee and whiskey—we are entertained by several tiny deer mice, which make lightning-quick inspections of all our clothes and gear. I take innumerable photos of places where a mouse had been a nanosecond before I pressed the shutter button.

After our bear experience at Cache Creek I'm a bit worried that I might get the eebies in our tent, but as we drift off to sleep even the howling of wolves seems comforting. I do have a nightmare, however. A young girl with long golden hair and a sprinkling of large, dark freckles is standing in the rain. The rain smells like tears, and becomes torrential. The girl slips and falls into the water and grasps frantically at the freestone riverbank as she is flushed ever farther downstream. I scurry down the bank and try to grab her hand, but she's always a tiny bit out of reach. My attempts to save her become ever more desperate; I have to run to keep up. A fluorescent pink Tasmanian devil with advanced cancer of the face tags along beside me, snapping at my ankles. My heart pounds in my chest, I can't breathe properly, I'm wet from rain or sweat or tears or blood. I make one last frantic attempt to grasp her flailing arm, but she is washed over the Upper Falls into the Grand Canyon of the Yellowstone. Frances appears at my side. "What are you doing?" she says, bewildered. "It's only a fish." I peer over the falls into a gin-clear mountain brook. And in a shallow pool I see that the girl was really a Yellowstone cutthroat trout, perfectly content.

The dream bothers me all night. Especially the bit about the fluorescent devil.

At sunup the air is perfectly still, as usual, but no fish disturb the flat-calm water. Perhaps it's because we are experiencing the hottest summer in decades.

It must be eight o'clock now and a few fish are beginning to feed at the surface, all well offshore. We use big hopper patterns in the hope that they will work as attractors, but to no avail. Finally we tie on smaller flies and longer leaders, and I hook—and lose—something big. Then the wind gets up and the rises stop.

We decide to walk to Beaver Creek, which flows into the lake midway along the northeastern shore of the main basin. Mathews and Molinero say it's a good-sized tributary containing lots of hungry cutthroats averaging sixteen inches. They also mention that anglers are dissuaded from fishing here because of the long walk in from the trailhead and the marshy conditions along the banks, so we expect to have the fishing to ourselves.

When we get to the northern shore of Heart Lake, at the point where the trail first hits the beach, we meet a park ranger. He asks us for our fishing licenses, but we have accidentally left them back in the tent. I offer to go and get them, but he makes a note of where we are camped and says he might check with us later in the evening.

He also points out wolf prints on the sand, and mentions that a few days ago he saw a sow bear and two cubs on a distant hill. I lead him down to the water's edge where I have just noticed a carp-like fish, bright yellow in death, with a downturned mouth. "A Utah sucker: *Catostomus ardens*," he says. "It's native to Heart Lake and can grow two feet long, six pounds or more."

Farther along the shore we see a couple of anglers polaroiding from the banks. From the way they are scanning and concentrating, we can tell that they are proficient. Soon we have caught up, and they stop to talk. "We're camped just a couple of hundred yards east of here on a grassy flat overlooking the lake," the older man says. "The fishing is a fair bit harder than we expected. The cutthroats are big but hard to see. Even when they rise, they're hard to fool." A bald eagle swoops toward the water, and we all think he's going to take a fish, but it's a false alarm. We agree that the fact the eagle is having a hard time makes us all feel better. We also agree that this makes us mean-spirited and absurd.

Frances and I end up polaroiding and prospecting a mile or more of Beaver Creek, but we encounter nothing. Maybe it's a spawning stream and the fish have dropped back to the lake. Maybe it's something to do with the drought—the "marshy" banks are hard as rock.

We arrive back at our camp by mid-afternoon, and while preparing snacks we notice a tree with two neat, round holes in the trunk, one six feet above the other. A squirrel sticks his head out of the top hole and then starts throwing stuff out—pine cone trash? He does this for half an hour, then starts filling up the bottom hole with what looks like fresh food—grass and pine cones? Suddenly we become aware that a lone mule deer has sauntered into camp. He is standing just ten yards from us and browsing his way down to the lakeshore. I carefully follow him with my camera all the way to the water's edge, where I see a muskrat swimming out to the drop-off. The muskrat dives to the bottom and bobs up a minute or so later with a bunch of leafy waterweeds trailing from his mouth.

He swims to shore, hurries up a grassy gutter, and disappears from view. A few moments later he reappears on the gravelly beach and swims back out to the drop-off. I watch him for half an hour as he gathers his stores, and by the time I get back to camp, Frances has made a fire and a cuppa.

I mention the muskrat, and also that there is a persistent strip of calm water along the sheltered southwestern corner of the lake. Frances says she's going to stay at camp and watch the wildlife. She insists that if I go fishing alone, I must take her canister of bear spray.

"You won't feel unsafe without it?"

"Not here in front of the fire. I'll worry about you if you don't take it. But don't dare lose it. I won't be able to enjoy the rest of the trip if you leave it sitting on the ground somewhere."

I clip the canister onto the small bum bag I use in lieu of a fly vest, and soon realize that I am much more scared of losing the bear spray than I am of encountering a bear.

At the exact point where the trail descends back to the water's edge, the calm water begins. Hallelujah! In the rocky shallows, between a distinct drop-off and the wooded banks, I can see half a dozen tailing cutthroats. Big ones. I look closer. They are sipping down caddisflies and spent BWOs (blue-winged olives). They are hard to polaroid, and as spooky as any brown trout I have ever stalked. I prick one, break off another, and finally land three, all about four pounds.

On the way back I think of how I have allowed my fishing to be adversely affected by the comments of the guys at the Parks' Fly Shop. In order to be able to see fish when visibility is difficult, you need to truly believe they are there. My attitude has changed now: tomorrow I will be able to polaroid fish no matter how hazy the conditions.

It's dusk by the time I get back to camp. Frances tells me all about the deer, eagles, and squirrels she has been watching. We drink wine. We study the night sky. We retire to our tent.

I can't sleep—perhaps I'm scared of having another nightmare—so I busy myself thinking about the history of the Snake River, of which Heart Lake is a headwater. I am looking forward to drifting the main stem with a professional outfitter in a couple of days' time.

The Snake River Indians are the Shoshone. In my childhood, the Northern and Eastern Shoshone fitted the bill as true Indians: they lived in tepees around the fringes of what is now Yellowstone National Park and hunted buffalo from horses.

The Western Shoshone, on the other hand, lived in Oregon and western Idaho on the Snake River below Shoshone Falls. This was below the range of the Yellowstone cutthroats in an area utilized by countless millions of spawning salmon, and the salmon were central to the tribe's life and culture. Indeed, they were known as the "Salmon-eaters." The name "Snake" Indians came from the S-like hand gesture they used to represent swimming salmon, which was misinterpreted by whites as a snake sign. The Snake River should really have been called the Salmon River.

The most renowned Salmon-eater was Sacagawea. She was abducted by Minnetaree Sioux when she was twelve, subsequently taken to an Indian village in North Dakota, sold as a "wife" to a French trapper, and ended up joining Lewis and Clark as a translator with their Corps of Discovery. She reminds me of Truganini, Tasmania's last tribally born Aboriginal woman, who also acted as a translator and mediator. Truganini was one of the very few aboriginal characters we learned about in primary school, but the tragedies and injustices of her life were largely ignored. Romance was made to trump reality.

The Snake River Shoshone were distressed by the encroachment of white settlers, viewing them as illegal immigrants who disrespected local culture, landscapes, water, animals, and trees. In the 1860s, gold prospectors began living on Shoshone land for longer and longer, exacerbating existing tensions, and from 1864 to 1868 there erupted a series of guerrilla skirmishes collectively known as the Snake War. The army was called in, and commander George Crook—the officer on the big white horse who carried a fine fishing rod—eventually, if belatedly, brokered peace with Chief Weahwewa.

Subsequently, the Shoshone fought alongside the US Army in the battles of the Rosebud and Little Bighorn, because it was waged against their enemies: against the ones who stole Sacagawea and also against the Lakota and Cheyenne. The alliance didn't help the Shoshone; in the end, the enemy of their enemy proved to be their enemy as well.

The next morning we break camp early. We encounter only three tent sites on the way back to the northern shore, and only one of these is occupied. On the face of it, the Yellowstone backcountry is manifestly overregulated and underused. On the other hand, I wonder if it can really be overregulated if there are fewer campers than campsites.

We arrive at the northern beach just before the eight o'clock rise. The sun is low and some smoke still lingers high in the sky, but we somehow manage to polaroid a few cutthroats even before the insects start hatching. Then, for the next two hours, we cast to dozens of giant risers, and even end up landing a few.

If this were a water at home, I would fish here week after week until I had observed and mastered all that I found interesting or elusive. For me, the biggest travesty of Yellowstone management is the expectation that a maximum of two pre-booked nights a year is enough to divine the inherent worth of a backcountry lake or stream.

My relationship with Tasmania's Western Lakes is more of a marriage than a fling or one-night stand. My bond to the land and water was built upon the progression of days, seasons, and years—nothing less could foster the sort of intimacy that makes me so protective of it.

In each of the Western Lakes there is a unique association between the quality of the spawning grounds, the availability of food, and the number and average size of the trout. The correlation is not static, but fluctuates according to rainfall and many other factors, and the effects of one good or bad year can take two, three, or more years to become evident. Furthermore, not all waters conform to the same part of the cycle in the same year. Over the decades, the

Greg French

most dedicated fly fishers become so attuned to the environment that they can predict events that others deem to be random: the years when most fish in any particular lake will be thin and "slabby"; the years when most will be relatively small but in their prime; the years when most will be at their heaviest. I wonder how long it would take for a casual visitor, or even a nonfishing researcher, to come up with the sort of accurately predictive reasoning developed by the best fly fishers in the Western Lakes.

On the walk back to our car we see where the sow bear and her cubs have been upturning and ripping apart logs in search of grubs. We also end up meeting some mounted park rangers (hell, there are more rangers than campers) and a guided party of anglers on horseback, their gear carried by accompanying mules.

Heart Lake has proved to be the quintessential American backcountry experience. But we are not done yet. From the Heart Lake trailhead we will drive on to Jackson, a quintessential American town, where we have rented a room for the night. And from here we will begin our quintessential American drift-boat experience down the Snake River.

SNAKE RIVER FINESPOTTED CUTTHROAT TROUT

The Snake River flows south out of Yellowstone National Park through Grand Teton National Park. It interests me because the section of river in Jackson Hole is home to the finespotted cutthroat (*Oncorhynchus clarki behnkei*). This is the only variety of cutthroat trout that completely dominates in its native range, and it is the mainstay of the local recreational fishery.

The finespot's status as a genuine subspecies is sometimes disputed because no genetic marker has been found that differentiates it from the standard Yellowstone cutthroat, but it has distinctive patterning and breeds true to type (offspring always look like their parents, never like other cutthroats). Furthermore, it boasts many physical and behavioral traits atypical of all other cutthroats. For a start, it is supposed to be an especially strong fighter, almost as good as the rainbow trout. Then it has a high resistance to whirling disease and (thus far) an unusual reluctance to interbreed with nonnative rainbows.

As Behnke points out, "There are no rules or standards of quantifiable genetic differentiation to qualify as a subspecies; only that a subspecies should possess one or more unique characters which differentiates it from all other subspecies." And in a country where so many unique races of salmonids have been watered down or obliterated through indiscriminate hatchery operations, the finespotted cutthroat stands out as a remarkable survivor.

Behnke even suggests that there is a reasonable case for classifying the finespotted cutthroat as a full species on the basis that it "maintains its identity by reproductive isolation from largespotted cutthroat in the upper Snake River system." Reproductive isolation, he stresses, is how other fish are separated as a species, including the pallid and shovelnose sturgeons, "which are indistinguishable by genetic analysis but have obvious morphological differences." He warns, however, that the finespotted cutthroat would then represent "an unusual evolutionary branching sequence: a subspecies, the Yellowstone cutthroat trout, giving rise to a full species." He is also concerned that no one knows the exact distribution of finespotted and largespotted cutthroat trout in the Snake system prior to the introduction of nonnative trout and the construction of artificial dams; historically, there may have been some areas of natural overlap and integration.

Behnke freely admits that his preference for the categorization of the finespotted cutthroat as a subspecies rather than a full species is a judgment call. He is, for example, in disagreement with many European taxonomists, who insist that Slovenia's marble trout is merely a subspecies of brown trout—another reminder, as if one was needed, that taxonomy remains a very messy business indeed.

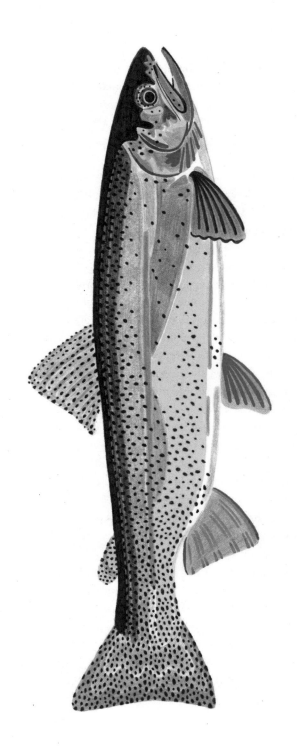

Snake River Finespotted Cutthroat Trout
(*Oncorhynchus clarki behnkei*)

The last part of the finespot's trinomial name, *behnkei*, was bestowed by popular writer M. R. Montgomery in his book *Many Rivers to Cross* (1995) as an acknowledgment of Behnke's detailed description of the fish in *Native Trout of Western North America* (1992). It seems strange to me that no formal name had been nominated prior to 1995, especially since the subspecies had been declared worthy of recognition as early as 1970. I also wonder why it was left to Behnke himself to formalize the name in scientific literature. (He was not being vain—there is a strict protocol in taxonomy that the first recorded taxonomic name must take precedence over any other.)

Quite besides this issue, I am mainly attracted to Jackson Hole because the local subspecies of the Yellowstone cutthroat looks distinctive. Really, though, why should the fact that it has the finest and most profuse spots of all native trout in western North America have any bearing on its importance to me or anyone else?

JACKSON

The valley known as Jackson Hole is defined by the Tetons to the west and the Gros Ventre Range to the east. With no pine forests to suffocate the views or vast prairies to alienate the foothills, the granite peaks don't merely dominate the scenery but call out to be explored. A siren song emanates from the chunks of unmelted ice on the immense Tetons and the ripples of golden sunlight on the vast Jackson Lake.

Suddenly we are being forced to stop. Road work. We are the first in the queue, which means we can expect to be delayed for up to thirty minutes. I begin to daydream, amusing myself with the coincidences of local etymology.

Jackson, Christian name Davey, was a pioneering beaver trapper, though the first white man to see the valley was probably John Colter, the white explorer who "discovered" Yellowstone Lake.

Hole was bestowed by early trappers, most of whom descended into the valley from high country to the north and east.

Teton is French for nipples, and it too is said to have been bestowed upon the mountains by the trappers.

Teton also happens to be a native Indian word and was used as a synonym for the Lakota Sioux.

Sioux is an abbreviation of Nadouessioux, which was a French approximation of an Ottawa Indian word long believed to mean "small rattlesnake."

The Shoshone, who hunted here in the Tetons and were the distant enemies of the Sioux, were known as the Snake Indians.

The road director comes over to my wound-down window and I say, "G'day."

"Aussies, eh?"

"Yeah. Spectacular land you got here, and you get to stand in it all day long. You must love your job."

"Sure do. I get to watch the whole world pass by. Today I've seen people from everywhere except maybe Luxembourg. Hey, you can drive on now. Have a nice day, won't you?"

Jackson is smaller than I expect, just 9,000 residents, and we quickly find the Anvil Motel, where we have booked a room. From our verandah there are pristine hills in all directions. They're close too; we're just a few hundred yards from the ski slopes of the Bridger-Teton National Forest.

We spread out our tent on the balustrade to dry, and our sleeping bags to air, and walk into town for a cold beer and a hot snack.

The shop fronts are reminiscent of buildings in the old westerns I used to watch on TV: verandahs shade the wooden sidewalks, and funnily enough, everything seems more quaint than kitsch.

We find a logical spot to cross a street but are arrested by a sign: "Whoa! Partner Use Crosswalk Down Yonder. Thank You." At the street corner there is a bucket of red flags. The idea, it seems, is that you hold a flag aloft and the traffic stops while you cross. Then you put the flag in a companion bucket on the other side.

Everyone is cheerful and down-to-earth—there are no suits, high heels, or makeup. Lots of shops display notices in Spanish as well as English, and a surprising number of people can be heard speaking Spanish. I ask a Spanish speaker about this, in Spanish, and he answers, in English, that the services in town are largely supplied by immigrants.

Some of the services are in restaurants and curio shops. In summer, additional services crop up: fly fishing, hiking, mountain climbing, kayaking, mountain biking, and horse riding. In winter there's skiing and snowboarding.

We walk on. A local sculptor has been commissioned to fashion bronze statues of the local wildlife—all life-sized and lifelike—and it seems that every park and esplanade is home to a pair of dueling elk, a patriotic bald eagle, or a stern bighorn ram.

Some towns usher tourists into their parks through topiary archways; here the archways are made of interwoven antlers. "Most are collected from the National Elk Refuge in winter shortly after the males shed them," says a helpful local. "Children can earn good pocket money scouring the countryside looking for them."

We pop into Wild West Designs and find innumerable elk-antler candelabras, elk-antler hat racks, elk-antler walking sticks. We can also purchase a string of Christmas lights made of 12-gauge shotgun shells, or T-shirts emblazoned with slogans like "PETA People Eating Tasty Animals." We ask a staffer if she knows where we might find an Internet café. "Everything's Wi-Fi," she says, and goes on to answer our other questions: "You can find a Laundromat on the next block. And you can get fishing licenses from the Orvis, Jack Dennis, or Snake River fly shops."

We end up getting a beer and sandwich from a bar. The front is open to the street, and on the sidewalk a bunch of local actors are performing the daily re-enactment of a supposedly typical scenario from the nineteenth century. They wear showy cowboy chaps and fake six-shooters, white hats and black hats.

We have another beer. In a nearby park a loud band starts singing in Spanish. One of the songs is "La Bamba" and I mention to the waitress that the band sounds a lot like Los Lobos, an old pop

group we used to hear on the radio when I was in my mid-twenties. "It *is* Los Lobos," she says proudly. "The original."

We walk home via the Rawhide Motel, Wagon Wheel Motel, Antler Inn, Cowboy Bar, and a shoe store called the Bootlegger. Strangest of all is the Jolly Jumbuck, a shop that sells "leather and shearling."

"How many Americans can possibly know what a jolly jumbuck is?" I wonder aloud.

"About the same number as Australians who know what shearling is," says Frances.

It seems that all ethnic groups have a stake in Jackson. All except indigenous Indians. Come to think of it, we've not seen one obvious Indian in all of America so far.

DIVERSITY WITHIN SUBSPECIES

Most Jackson locals use drift boats, and it was because I couldn't find a drift boat for rent, that I opted to hire a guide. We chose Dave Barber more or less at random from the Internet, and booked months in advance. His postal address is in Jackson, which seems perfect, and he has assured me that we will catch some Snake River finespotted cutthroats, probably lots.

Last night I rang him to confirm the trip, and he gave me directions. "Take the main road out of town..." He went on to mention a series of turns and overpasses and gave helpful hints like, "If you reach the bridge over the so and so, you've gone too far." I retained what I thought were the essentials: Highway 189; right at Hoback Junction onto Highway 89; will see you on right-hand side of road outside the Etna Trading Company at 9:00 a.m.

The problem is, it's 8:50 a.m. now, and we are hopelessly lost. I manage to buy a detailed map from an outdoor shop, and am

Greg French

stunned to discover that Etna is more than fifty miles away, an hour's drive at least. Also, it's nowhere near the Snake River.

Frances phones Dave, confirms our suspicions, says we'll be late.

There's plenty of interesting stuff to see en route—the gorges of the Snake Valley, the rural flats beyond the gorges, turkey vultures nesting on transmission poles, bald eagles playing in the thermals—but my mind starts to wander and soon I'm thinking about the differences between the regular "large spotted" and the "Snake River finespotted" varieties of Yellowstone cutthroat trout.

Interbreeding between different "full" species, such as occurs between rainbows and cutthroats in the Lamar catchment and lower Yellowstone River, is one thing, but interbreeding between different subspecies and even races of the one species is just as problematic.

By the early 1900s, American anglers had become acutely aware that sea-run rainbows were not only different from river-resident stock, but different among themselves. They varied in the time of year that they ran from the ocean into freshwater, in the length of time they spent in the river before spawning, in the dates they spawned, in the time smolts (juveniles) spent in the river before going to sea, and in the number of years the adults spent at sea.

We now know that there is even more to the story. Quite apart from migratory differences, some trout use different parts of a river to spawn, some are better adapted to fast water or slow water, some prefer cold water or slightly warmer water, and some do best in clear water or glacial water. Furthermore, some are specially adapted to feed on food specific to certain parts of a river or lake.

In the earliest days of angling literature, most Americans of European descent suspected that these traits were mainly influenced by environmental conditions, or that they occurred randomly, but by the late 1800s some anglers and naturalists were wondering if they might be inherited. Unfortunately, the leading scientific authorities of the era pooh-poohed such ideas, insisting that all the rainbows in any river mixed randomly on the spawning beds.

With the benefit of hindsight we now know that these early scientists were mistaken. By the 1970s, plenty of experimental evidence showed that, in the wild, resident rainbow trout give rise to predominantly resident fish, steelhead to steelhead, winter-run fish to winter-run fish, summer-run fish to summer-run fish. In some studies the degree of separation has been shown to be very close to 100 percent. Often in such cases, the search for specific genes that might code for specific migratory behavior has been elusive, but Behnke points out that only a tiny sample of genetic material is examined in any study. No one is suggesting complete reproductive isolation between fish with differing life histories, but clearly the isolation is profound, certainly sufficient to maintain the integrity of the hereditary basis that separates one race from another.

The truth is, genetic coding for homing and other reproductive traits is much stronger and much more specific than anyone from the nineteenth century could possibly have imagined. So too is the extent to which adaptive life history is related to the abundance of wild trout stocks.

It's easy today to judge too harshly the nineteenth-century scientists who thought about heredity and drew the wrong conclusions. Our anger should really be directed toward the people, including hatchery managers, who didn't stop to think at all.

Rainbow trout were first used in artificial propagation in 1870, and were sourced from streams running directly into the southern part of San Francisco Bay. By 1877, eggs were also being collected from river-resident redband trout in the McCloud River, a distant headwater tributary of the Sacramento River, which drains into the northern part of San Francisco Bay. From 1880 to 1888, hatchery staff were mixing these fish with steelhead, probably the Central Valley form of coastal rainbow trout. After 1888, the US Fish and Wildlife Service added steelhead from northern California and southern Oregon.

According to Behnke, the forced mixing of various races of rainbow trout during artificial propagation "acted to break down the reproductive isolation that is necessary for wild populations to maintain adaptive life history traits." In short, it destroyed population diversity.

Behnke also notes that the genetic variability of the founding domestic stock allowed for "rapid domestication, based on such factors as efficient utilization of artificial diet, rapid growth, early sexual maturation, and high fecundity. Domestication is also based on artificial selection for behavioral modifications, such as the ability to tolerate crowding and a willingness to come close to humans for feeding." Furthermore, Behnke stresses that all hereditary changes in hatcheries run counter to natural selection, where the sole criterion is "survival to reproduction in the wild."

Does this matter?

Absolutely.

Today in the Columbia River catchment—arguably America's most famous trout and salmon producer—75 percent of all steelhead returning to the river from the ocean are hatchery stock that were released as smolts, and Behnke says that pure populations of steelhead uncontaminated by hatchery rainbow trout are now rare. The domestics and the bastardized wild fish don't breed very successfully—they just don't know when to run, which tributary to run to, or where to spawn—so without ongoing artificial production, fish numbers, already miniscule by historic standards, would crash even further.

Maintaining big runs of fish through hatchery releases is extremely expensive, and most anglers would prefer wild fish anyway, so what can be done? One idea was to add fish ladders to the dams and mandate environmental flows so that any remaining wild stocks could find their way home. Unfortunately, this did little to improve the abundance of wild stocks; most of the original genetic strains—those programmed to make the best use of each section of river—had already become extinct or were continuing to crossbreed with hatchery fish.

The next idea was to select hatchery brood fish from wild spawners in the parts of the river system where the offspring would be released (usually at smolt stage). Over the last couple of decades this has been tried in various rivers on Vancouver Island (British Columbia), where the steelhead runs have been decimated as a result of poor logging practices. But the evidence from Vancouver is that, even when you rehabilitate the environment, it is depressingly difficult to re-establish steelhead runs by relying on hatchery stock from *any* source.

The burning question became: *How quickly can domestication occur?*

Most researchers suspected that it would take multiple generations to see genetic evidence of domestication, but the authors of an article published in the *Proceedings of the National Academy of Sciences* in December 2011 found that the impact of hatcheries is so profound that after a single generation, hatchery fish become genetically distinct from wild fish and have difficulty surviving in the wild. It is not yet known which genetic trait is being (unintentionally) selected for, but it is likely to be linked to an ability to withstand overcrowding.

In hindsight, these sorts of problems were probably inevitable. When eggs are taken from the wild to a hatchery and the hatchlings on-grown to smolt size, too many fish survive—including almost all of the ones that nature would rather weed out. If the eggs were left to hatch in the river, only a tiny percentage of the resulting fish would reach smolt stage, and these would be the ones best suited to life in the wild.

So how does all this affect management strategies for the land-locked cutthroats in Yellowstone Lake? Well, the reason that managers are so reluctant to bolster the lake's greatly depleted stocks with hatchery-reared fish is that, based on the experience elsewhere, they would run the risk of reducing genetic diversity and also of domesticating the stock, either of which could ultimately result in extinction.

Greg French

The genetic diversity in Yellowstone Lake's cutthroat trout is not well understood, but the situation at Pelican Creek, Clear Creek, and other lake tributaries suggests not only that adult fish are strongly inclined to spawn in their natal rivers but that the homing instinct is genetically ingrained.

There is also a good chance that disease-resistant races exist somewhere in the Yellowstone Lake, and if large numbers of hatchery fish were released into the system, any such strain could be overwhelmed.

It depresses me that hatchery managers on mainland Australia produce countless numbers of native fish—Murray cod, Australian bass, golden perch, you name it—and distribute them willy-nilly all over the place. I am definitely coming to believe that we should be mimicking the cautious approach being followed by fishery managers here in the Greater Yellowstone Ecosystem.

SALT RIVER

The native range of the finespotted cutthroat trout comprises the main stem of the Snake River from the Jackson Lake Dam downstream to the Palisades Dam, including all associated tributaries except for the first three creeks downstream of Jackson Lake (Pacific, Buffalo, and Spread). No physical barriers isolate the finespotted cutthroat from the largespotted variety, yet the species' distribution shows no overlap. Behnke surmises that shortly after the last glacial peak some 20,000 years ago, a series of landslides or ice dams temporarily isolated a population of largespotted trout that subsequently evolved into finespotted trout. When the dams eroded, the finespotted trout were able to outcompete the largespotted ones in big-river environments, in part because they were used to living alongside a wide variety of native fish including suckers, sculpins, and minnows.

The town of Etna, we finally discover, comprises just a handful of buildings, and we soon spot a four-wheel drive hitched to a McKenzie-style drift boat outside the Trading Company. "You must be Dave," I say as I get out of our car. He nods cheerfully and says we should gather our gear and hop into his car pronto. "The morning's getting old," he tells us.

Dave is a friendly family guy and pokes fun at us for getting lost. When I ask, he says we'll be drifting the Salt River. I look it up on the map and am relieved to see that it runs into the Palisades Reservoir. We will be fishing finespot habitat, after all.

The place where we launch reminds me of the Macquarie River back in Tasmania: a faintly murky broadwater flanked by flat expanses of straw-colored pasture. The current is slow but distinct, and I can see small fish rising here and there in the bubble lines running adjacent to the undercut banks.

Surprised by the parched landscape, I ask Dave whether he thinks climate change is having much of an effect on local weather patterns.

"Do you think climate change is real?" he says.

His tone is neutral, but I sense skepticism and tell him all I have witnessed: the dieback of vast swaths of British Columbia's conifer forests; the shrinking and demise of icecaps and glaciers in New Zealand and South America; unprecedented floods, droughts, and firestorms in Australia; catastrophic tidal surges on Pacific atolls—all predicted by climatologists.

"But from what I've heard, not all scientists agree that the stuff you talk about is caused by climate change," he says.

"Actually, for all intents and purposes, they do."

"I guess there's so much contradictory stuff in the media, it's hard for many folks to know what to believe."

"Maybe it's not what you believe but who you believe," I suggest. "If you have a heart attack and your doctor says you need a bypass, you might get a second opinion from another doctor—after all, some professionals are better than others—but would you prefer the advice of your local mechanic? With climate change, the peer-reviewed research is unequivocal. Why seek a second opinion from a hack columnist in the local newspaper? Or from an apologist for the fossil fuel industry?"

"But no one is certain about climate science, are they?"

"In science, nothing is certain. No scientist will claim with certainty that you will die if you fall out of a plane at 30,000 feet. It's all about probability."

He nods.

"Look at it this way," I continue. "What would you pay *not* to put your loved ones on a flight that had, say, a 10 percent chance of falling out of the sky? Most experts assess the risk of planet Earth becoming unimaginably inhospitable for humans to be more than 95 percent. Many put the figure at very close to 100 percent. Sure, the planet has been warmer than it is now, but only in the geological past. The Jurassic climate was good for giant lizards, not people."

Dave listens with apparent interest, and is respectful, but I realize I have been ranting and feel I should change the subject. "What flies will we be using?"

"Snake River cuts feed well on whatever is available at the time—everything from decent-sized fish to tiny insects and crustaceans, regardless of whether the food lives on the bottom, in mid-water, or on the surface. All cutthroat trout do this to some degree, but the finespotted variety is king."

"Do you catch any nonnative trout in finespot habitat?"

"In the Snake and tributaries like the Salt, we occasionally catch a brown trout or even a brook trout, but only occasionally. I've also heard of the odd mackinaw—probably migrants from Jackson Lake."

"Rainbows?"

"There's rainbows in the Snake River catchment for sure. Emma Matilda Lake and Two Ocean Lake have got them. I think there's a few in Jackson Lake too, though people mainly catch mackinaw, cuts, and browns. They even catch one or two rainbows in the Gros Ventre River."

"The Palisades Reservoir?"

"It's mainly native finespotted cutthroat and some large browns. There are a few mackinaw and nonnative kokanee, but I've not heard of any bows."

"So what fly should I tie on today?"

"Today? Well, today is a hopper day."

That's handy, I think, because that's what we've got on. But Dave doesn't like our neutrally buoyant deer-hair flies, the ones that slide into the mouths of the sippiest trout. He thinks we should use hoppers with big foam bodies that ride high on the surface. He doesn't like my Maxima tippet either—"Thick as rope"—and he replaces it with one of those trendy, small-diameter brands that have less-than-optimal knot strength.

I ask if he has any kids, and he says he has two young sons.

"What are their interests?"

"You know, they're good readers and they like fantasy."

"Cool," says Frances. "Have they tried *The Hobbit* or the Harry Potter books?"

"Well now, we don't really hold to that sort of stuff. I haven't read them myself, you understand, but from what other people say, I don't think you can be too careful about the sorts of things young minds are exposed to."

A few small fish have a go at our flies but can't seem to suck them down. Then I hook a big one, and it breaks me off. Oh well, it's probably best not to hook too many tiny trout, and maybe it would be good for me to practice being more gentle on the strike.

Soon enough, Frances and I have each boated a few finespots, all about one pound, and by early afternoon we have each landed a couple in excess of two pounds.

By now Dave has sussed out our personalities and has deftly moved the discussion onto topics we all find enjoyable. "Did you know that all schoolchildren here in Wyoming's Lincoln County are granted a day off school for the opening of the elk-hunting season? And they can take another twenty random days off in a

Greg French

year, too, providing their parents assure the school they are doing outdoor stuff."

Frances and I agree that this is a good thing.

"It's why I moved here from the city. Kids need to learn how to be self-reliant. You know, we hardly buy meat anymore. Our freezer is full of home-shot elk."

Some of Dave's fishing strategies seem strange to me—like persisting with small recalcitrant risers when I think the time might be better spent polaroiding bigger, bolder fish—but there's no denying that we're having a good day. I also like the way he takes time to show us beaver lodges and osprey nests, and to talk about local culture and history.

Lunch is a banquet. Dave sets up a picnic table on the riverbank and perfectly fries some enormous pork chops, which he serves with salad and a choice of quality beer or wine, all locally produced. He has proven himself to be an engaging and humble guide of considerable competence.

After lunch we catch more finespotted cutthroat trout, but at about four o'clock a huge rainstorm ambles in from the mountains. The lightning gets menacingly close and we agree to call it a day, but not before I land one more trout, a large one with stunning golden hues on its head, brilliant red slashes on its jaws, and especially fine spots on its body. This fish has me wondering what biologists would make of the Snake River cutthroat were it not *visibly* different from the standard largespotted variety. Would they think that its unique behavioral traits were merely the result of environmental factors rather than hereditary ones? Would they notice the behavioral traits at all?

Genetically distinct races of cutthroat trout could easily exist in Yellowstone Lake, even if we can't tell them apart visually. I wonder: *What inherited traits are going to prove most important in the face of unprecedented threats from climate change, nonnative fish, exotic diseases, and other perils as yet unimagined?*

Chapter 9:
Back to Bozeman

FIREHOLE

Before we leave the Tetons, we spend a day canoeing on Jenny Lake and Leigh Lake. Unlike anywhere in Yellowstone National Park, this area teems with day walkers. I suppose the difference is that the most attractive features—the dramatic mountains and intimate lakes—are instantly available; you don't have to walk across miles of open grassland to get there.

Then we spend another day on the upper Gros Ventre River—above the Lower Slide Lake in the vicinity of Crystal Creek, northeast of Jackson—where we explore the semi-arid landscape in solitude and catch plenty of finespotted cuts.

Now it's time to make our way back to Bozeman, and we are going to go via the Firehole.

The Firehole River, a tributary of the Madison, is flanked by geysers, including Old Faithful. Since childhood I have been inspired by photos of anglers fishing meadow-banked riffles and weedy glides amid sulfur, steam, and buffalo.

According to my guidebooks and the park website, originally there were no trout at all in the twenty-odd miles of the Firehole River above Firehole Falls (less than one mile upstream of the Madison Junction). Brook trout were liberated in 1889, browns in 1890, and rainbows in 1923, and the latter two species thrived, providing world-renowned opportunities for serious fly fishers.

Modern environmentalists are prone to pooh-pooh anglers who favor difficult-to-catch nonnative fish in the Firehole over easy-to-catch native fish in the Yellowstone River downstream of Yellowstone

Lake. It is elitist snobbery, they insist, apparently unaware that this is the same argument used by proponents of domestic hatchery-reared fish over those who champion wild, river-born fish. Of course, the serious angler's attitude has nothing to do with snobbery and everything to do with the pursuit of variety and excellence. Familiar, easy fishing soon becomes boring. Dynamic, difficult fishing forces you to lift your game, and remains exciting for as long as it continues to offer new challenges and imperatives to learn.

Alas, we hear on the radio that the latest heat wave has resulted in the closing of fishing in the Firehole until further notice, lest the fish released by anglers die of stress.

"Well, we might as well go that way home anyway," Frances says. "At least we'll be able to say we've seen the Old Faithful Geyser."

Frankly I'm more interested in the Old Faithful Inn, and the chance to down a few cold beers is only part of it. Before we left Tasmania, a friend loaned us a copy of *Free Fire*—a thriller by C.J. Box where the action takes place entirely in and around Yellowstone—which offered an enticing description of the building's dramatic history and quirky architecture.

The Grand Loop Road runs down the lower two-thirds of the hole (valley) for which the Firehole River was named. We pull into the expansive tarmac of the Old Faithful car park, and are surprised by the immensity of the hotel, then surprised again by a spectacular eruption from the geyser. By the time we get out of the car, the eruption has finished and countless thousands of tourists are returning to their vehicles. We decide to dodge this tsunami of humanity by delaying a look at the inn and exploring the boardwalks that interlace the Upper Geyser Basin.

People are already gathering for the next eruption, due in about an hour. We walk past innumerable fumaroles, hot springs, and mud volcanoes and soon cross the Firehole River, which, like everything else, is steaming. The most renowned fishing occurs downstream, below the coldwater input of the Little Firehole River. Even so, summer temperatures in the main part of the fishery often become hot enough to send brown trout into torpor and to kill rainbows.

According to my guidebooks, the geothermal inputs do other strange things to the Firehole River. All wild rainbows are predisposed to spawn when temperatures are rising, and this usually happens in the springtime. Here in the Firehole many rainbows traditionally seek refuge from the summer heat by migrating into the coldwater tributaries, but they drop back as soon as they can, and therefore commonly experience the requisite temperature increase in the fall.

The highly mineralized water also bolsters biological production and accounts for the large size of fish mentioned in historical literature. Mind you, these days the fish caught by anglers are said to average less than ten inches (maybe half a pound) and only a few reach twenty inches (three pounds or better).

The problem, I am told by almost everyone who fishes here, is climate change. Summers are staying hotter for longer, and when temperatures in the main stem approach lethal highs, abnormally high numbers of trout migrate into the tributaries. This results in disrupted feeding patterns, both for the refugees and the residents, and the sudden increase in population density among already-stressed fish becomes a recipe for disease.

Everyone agrees that the fishing in the Firehole is still very worthwhile most of the time, but they all worry about its future.

By the time we get back to Old Faithful Geyser, thousands of people are sitting in an amphitheater of wooden platforms, waiting for an imminent eruption. The social spectacle is at least as interesting as the geothermal one, but we decide to race over to the hotel while the crowd is preoccupied.

The lower story of the Old Faithful Inn is constructed of lodgepole pine logs, but is completely overwhelmed by the upper stories, which are clad with timber shingles and mostly contained within giant, steeply pitched gables. According to Wikipedia, it was constructed in the winter of 1903–04, "the first of the great park lodges

Greg French

of the American West," and is possibly the largest log building in the world, certainly the largest log hotel.

We walk inside. The lobby is a vast open space, seven stories high from floor to ridge, centered around a massive stone fireplace and rimmed by layer upon layer of balconies and stairways, all supported by natural logs braced with quaintly twisted limbs. The framework seems ridiculously insubstantial—as if it supports itself only through the will of its creators—and radiates an elfin charm reminiscent of Caras Galadhon, the City of the Trees in Tolkien's *Fellowship of the Ring*. I am a builder by trade, and I can scarcely accept the ambitiousness of the architect or the skill of the craftsmen. The organic materials seem wild and alive, like the forests and animals outside. Truly, the preservation of constructed environments can be as important as the preservation of natural ones.

LAHONTAN TROUT

We have arrived back at Bob and Karin's house in Bozeman. Unfortunately they are stranded in Santa Fe, their flight home having been cancelled due to a technical fault with the plane. Oh well, at least I'll have time to finish my homework.

The day after tomorrow we will fly to Colorado and drive from Denver to Fort Collins to interview Dr. Robert J. Behnke, and I have one more chapter to reread in *About Trout*, the one titled "Were Fish Really Bigger in the Old Days?" It starts off talking about Nevada's state fish, the Lahontan cutthroat trout (*Oncorhynchus clarki henshawi*).

I've caught coastal cutthroats on Vancouver Island, westslope cutthroats in the British Columbian Rockies, and Yellowstone cutthroats in Yellowstone National Park, and when planning this trip to the United States, I had originally hoped to make a clean sweep of the four ancient cutthroat lineages by taking time to visit Pyramid Lake. I find the Lahontan trout especially appealing because it is the most sharply differentiated subspecies of cutthroat, having been isolated in the Lahontan basin for almost a million years.

Pyramid Lake is the largest natural still water in Nevada. In fact, at thirty miles long by more than eight wide (almost 200 square miles), it is the second-largest natural lake in the western United States. It and neighboring Walker Lake (eighteen miles by seven) used to be part of ancient Lake Lahontan, which 12,700 years ago covered some 8,500 square miles. These days both waters are located in a dry, dusty salt lick and owe their existence to snowy streams flowing off the Sierra Nevada. Temperatures are very hot in high summer and very cold in deep winter.

For several reasons I ended up not wanting to go to Pyramid Lake. For a start, it turns out that there are no wild fish left—the famous Pyramid fishery is now completely maintained by hatchery releases. Then I found out that shore-based angling is mostly done along the edges of drop-offs in chest-deep water from atop portable stepladders.

I might still have gone there but was mercifully let off the hook; the only reliable fly fishing occurs in March and April when the cutthroat come out of the depths in order to spawn on the shoreline shelves, a desperate act that results in no natural recruitment. Right now it is high summer—the shoreline shallows will have reached intolerable temperatures and the cutthroat will have retreated back below the thermocline.

According to Behnke, as Lake Lahontan shrank, salinity and alkalinity increased to a point that makes the water intolerable to most salmonids, yet the local cutthroat trout did not merely adapt, they thrived.

Historically, typical Pyramid cuts lived for ten or more years, and grew huge. I've seen a sepia photo from 1923 by Wallace Upson showing a typical catch of perhaps twenty twenty-pounders. Behnke notes that the official record for a rod-caught fish is forty-one pounds, but acknowledges that rare specimens may well have grown to over sixty.

Greg French

For thousands of years, native Americans caught the Lahontan trout, dried them on racks, and ate them in the cold of winter when the lake was frozen over. Then in 1858 the Comstock Lode was discovered under what is now Virginia City. Miners flooded in from California and farther afield, and all had to be fed. Soon, commercial fishing was under way at Pyramid Lake. The market developed to the point where Pyramid fish were being sold as far away as Salt Lake City and San Francisco. This was, of course, unsustainable.

Then, in 1905 a dam was completed on the Truckee, blocking access to two-thirds of the lake's spawning grounds. Increasing draw-offs meant ever less water in the lower third, and by 1943 the Pyramid strain of the Lahontan cutthroat was extinct. (By the 1950s the entire lineage was thought to be extinct in the wild throughout its native range.)

In the late 1940s, the State Fish and Game Commission released kokanee, rainbows, and domestic Lahontan cutthroats into Pyramid Lake, with limited success. In 1960, they trialed rainbow and cutthroat hybrids (cutbows) with somewhat better success.

Then Behnke arrived on the scene. He felt there must still be some wild Lahontan cutthroats above barrier falls on headwater creeks, and soon discovered native populations in the headwaters of the East Carson River. Another natural population was subsequently discovered in Summit Lake.

In 1974, management of Pyramid Lake was transferred to the Paiute, who were determined to re-establish a pure Lahontan cutthroat fishery using wild stock from Summit Lake.

The Paiute sought to maintain "wild" status by collecting eggs and milt from the dense concentrations of spawners that soon began migrating along an expansive beach. Later they built an artificial spawning channel, into which they pumped water from the lake. The fish continue to run up this channel every March and April, where they are trapped and stripped. In order to permit natural selection, hatchery-reared fish are mostly released at fingerling size—the fish that survive to maturity in the lake are those that are incrementally adapting to changes in lake chemistry brought about by shifting water levels.

Yet the Summit Lake fish have never grown as big as the native stock; most weigh three to four pounds and ten-pounders are uncommon.

Behnke came to believe that the original Pyramid Lake fish must have been genetically distinct. He also believed that there was a good chance of finding a genetically pure translocated population, perhaps in some headwater creek. And he was adamant that if returned to their home lake, these tiny creek-fish would grow as big as their forebears. In the 1970s, this line of thought was revolutionary. Many considered it crackpot.

The reason Behnke was confident of finding translocated fish was that a successful private hatchery had been established on the Truckee River in 1867 or 1868—possibly the first time any species other than brook trout had been propagated in North America—and fry and fingerlings had been widely distributed.

In the late 1970s, an unusual-looking trout was discovered in Nevada's remote Pilot Peak and Behnke—by that time the world's leading expert on salmonids—declared them to be genuine Pyramid Lake trout, basing his decision on appearance, the "most plausible" genealogy, and gut feeling, though no genetic testing has ever confirmed this.

However, Behnke still seems confident with his original conviction and maintains that DNA evaluation is heavily overvalued, that a meaningful understanding of species and speciation can only be gained by interacting with fish and their environments. For him, trout remain living, spiritual beings; he feels that if you don't have a spiritual affinity with them, you can't really appreciate them at all. "Look at what you can see," he insists.

Not everyone was enthusiastic about returning Pilot Peak fish to Pyramid Lake. The Paiute had a long history of success with the Summit Lake stock and were reluctant to tamper with a good thing. Furthermore, in order to ensure that Pilot Peak fish could never hybridize with Summit Lake fish, the brood stock would have to be forever maintained in an isolated hatchery. Steps could be undertaken to guarantee genetic diversity, but *natural* selection would not occur.

Nonetheless, in 2006 an agreement was finally reached: the Paiute would continue to manage Summit Lake fish while the US Fish and Wildlife Service would release a smaller number of Pilot Peak fish.

Six years have passed. Speculation about the origins of the Pilot Peak fish, or at least about how big they will grow in Pyramid Lake, will soon be overtaken by events. If Behnke is the guru so many people think he is—if he is right about the origins of Pilot Peak stock and right about the importance of unquantifiable genetics—twenty-pound cutthroats should be taken from Pyramid Lake this very year.

BOB AND KARIN

Our perfect hosts have arrived home. We sit on the deck of Bob's wonderfully designed home eating Karin's wonderful homemade huckleberry muffins, and after enjoying a few stories about their time in Santa Fe, we begin to reminisce about our first meeting in Mongolia in September 2010. We had gone there to catch taimen.

Taimen (*Hucho hucho taimen*), or "tuul" in Mongolian, are the world's largest salmonid and one of the most ancient. Russians call them Siberian salmon, and Bob and Karin liken them to steelhead. I'm tempted to tell you that they resemble elongate brown trout. Others have compared them to sharks, snakes, even dinosaurs. In truth, they defy classification—they are unique, mystical, and utterly captivating. They much prefer fast rivers to lakes. They commonly live more than fifty years. They are said to reach a length of six feet and a weight of 220 pounds. (This is probably an exaggeration, but plenty of validated reports have recorded fish measuring around five feet and more than 120 pounds.) Up to 70 percent of a taimen's diet is fish, but they have a creepy penchant for things warm-blooded—mice, squirrels, voles, even marmots.

Deciding which river to fish had been enormously difficult for me. Most of Mongolia's famous fly-fishing destinations—including the Delgermörön, Egiyn, Üür, Selenge, Ider, and Shishkhed—flow north into the Yenisei River and the Arctic Ocean. Yet I became fascinated with the biologically diverse Onon—a tributary of the Amur, which flows east before entering the Sea of Okhotsk near Sakhalin Island. Here, in addition to the taimen and the ubiquitous lenok (or "sharp-nosed lenok," *Brachymystax lenok*), I might be able to catch the rare Amur trout (or "blunt-nosed lenok," *B. savinovi*).

I had hoped to plot my own trip, but the logistics and economics of organizing rafts, fishing permits, translators, and other essentials became overwhelming. So I eventually swallowed my pride and arranged for a trip with Mongolia River Outfitters (MRO), selecting this business from among other reputed outfitters because it was the only one that operated on the Onon.

Bob and Karin, on the other hand, had chosen MRO because the man who set up the operation—Mark Johnstad—lived in Montana, just up the road from their own home.

Tonight we two couples delight in reminiscing about our chartered flight aboard a Cessna Caravan from Ulaanbaatar (UB) to the remote rural outpost of Binder. "Do you remember," says Karin, "how when we landed on the open steppe, people came from far and wide to witness the event; how we found ourselves surrounded by motorbikes, horses, and those Russian four-wheel-drive Kombi-style vans?" How could we forget it—the welcome had been as surreal as it was sincere.

We also fondly recalled our guides—Jaime, Daniel, and Fabián from Chile; Peter from Aruba; and Yuruult from Mongolia.

We agreed that drifting down the river had been fantastic fun. The fact that MRO used rafts rather than jet boats had been a major plus in my book: rafting is peaceful and unobtrusive, and seemed respectful of Mongolian culture. Really, though, I was fully aware that rubber rafts had no place in Mongolian history, and I still wonder if I was complicit in helping to make the Mongolian people ever so slightly more Western. I also can't help but wonder if Mongolia will end up making exactly the same environmental mistakes that America has made; capitalism seems to bring with it certain predictable consequences.

Frances quickly turns the conversation back to the good times, and Karin admits to a love of the gers. "The hand-painted framework was adorable. I loved the lattice walls and the numerous spoke-like rafters, the way the framing was covered with sheep's-wool felt. They were so cozy inside and, because of the hole at the apex, much brighter than I thought they would be, at least when the sun was out. The whole feel was very much like a tepee."

I loved the gers too, and remember saying as much to Bolormaa, our camp manager and translator. She responded, "I always hear fishermen say that they prefer natural rods to artificial ones, that cane is good for the soul. In the same way, gers are good for the soul." She seemed to divine our distaste for UB. "Russian buildings are full of harsh lines. Shadows of sadness and unease dwell in every corner."

Bob remembers someone telling him that Bolormaa was born to a golden lineage, that she could trace a direct line of ancestry all the way back to Chinggis (Genghis) Khan. He also remembers the evening I bragged about my five biggest lenok, all well over three pounds, all polaroided. "I trumped you with my Amur trout story, didn't I? That fish weighed more than five pounds. Sure, I snapped the butt of my rod while fighting it, but hey, there's plenty worse ways to break gear."

Frances recalls our first taimen, a small one. "In my hand it seemed mythical, like a baby dragon. Peter said that it might live for another half-century, that it could conceivably be caught a thousand times over. No wonder our Lonely Planet guide insisted that killing a taimen would bring misery to 999 souls."

As for the first big taimen, all I remember of it was the sudden weight and fury of a giant, miraculous animal. Of fighting it and Jaime netting it and all of us yelling in delight and touching the skin and setting it free. And sitting exhausted in the boat, bewildered, my endless thank-you's blurring into an "om."

I recall, too, the time when lenok began rising to midges. They had been difficult to polaroid in the glare and difficult to fool, so we had rowed among them to get a better look. Never had I seen so many trout crammed together, not even during a spawning run. The school was a hundred yards long and up to five wide, and densely packed. I had the feeling that this must have been exactly what the cutthroat trout had been like in the Yellowstone River in Custer's day.

We all agree that the local customs were an essential part of our Mongolian experience. Each evening, after the main meal, some nearby nomads would supply us with distilled milk vodka, and Bolormaa would lead us in accepting the offering. "Dip your left ring finger into the cup, flick droplets to the four cardinal directions—eternal sky, mother earth, friends to the left and right—then wipe the finger across your forehead, to bless yourself."

Frances remembers lying down on the grass at night and looking up at the heavens. "The starry sky was so vast and silent that I swear I could hear angelic arias, encouragements from Bolormaa's forebears."

Finally I recall the last morning, immediately before being driven to our plane, watching forlornly as a formation of demoiselle cranes flew south, gentle gray against vivid, endless blue. In the silence, I became aware of Bolormaa at my side. "In Mongolia it is bad luck to watch the cranes as they leave," she had offered. "We prefer to celebrate their return."

And return I will, in less than two months. This time I will be thinking a lot more about whether it is inevitable that Mongolia makes the same mistakes on the Onon in the early 2000s that America made on the Yellowstone in the late 1800s.

Greg French

DAN BAILEY'S
FLY SHOP

I am in Livingston, standing outside Dan Bailey's fly shop, "established in 1938," looking up at three giant—not to mention garish—fiberglass fish: a brown, a bow, and a cut. The façade, not quite crumbling, has an Art Deco feel to it. I feel at home: there are entire towns in New Zealand that, having been destroyed by earthquakes in the 1930s, were rebuilt entirely in the then-current Art Deco style.

Normally, I have no interest in fly shops. For me, fishing mostly revolves around trout and the environments in which they live. I am interested in this fly shop because it was featured in Brooks's *Trout Fishing* and also in the 1975 film *Rancho Deluxe* (in which the lover of one of the protagonists worked here tying flies). Tasmania in the 1970s had hardly any specialist tackle shops, let alone fly shops, and the whole concept seemed highly exotic.

I was surprised to learn from Bob and Karin that Dan Bailey's still existed—Bailey died in 1982, ten years after *Trout Fishing* was published, but his son John still runs the place. I am going inside for the same reason that I fish the Tongariro River and the Rotorua Lakes: history and culture matter. I intend to buy some saddle feathers, and I am thinking about counting coup.

Among the artifacts found at the Battle of the Little Bighorn were coup sticks that belonged to Crazy Horse and Sitting Bull's braves. The Plains Indians won prestige not only by striking a blow to the enemy but also by other acts of bravery, including touching an enemy combatant with a coup stick and escaping unharmed. Coups were counted by putting notches in the stick, or by tying an eagle feather to it. The feathers could be notched too, or clipped or dyed to signify specific coups or other exploits, and some were worn in the hair.

I carry in my mind, all the way from childhood, detailed images of Crazy Horse, Sitting Bull, and other men, and I remain

especially fascinated by the way they wore their feathers with dignity and respect.

I have no time for the concept of the Noble Savage. To me it is an absurd idea that in a state of nature humans are essentially good. People are people, nothing more and nothing less. So why have I always been on the Indians' side? At age fifty I admire their deep love, understanding, and respect for the natural world. Australia's Tim Flannery, author of *The Future Eaters*, would argue that this respect was a reaction to *having* to live the way they did and of *having* to make their traditions work, not due to any inherent wisdom or cultural superiority, and I would agree. But it doesn't alter the fact that the respect was there. Why did Custer's image scare me witless? Because I felt, in the way only a child can feel, that the eyes and demeanor were callous; to me, Custer exuded disrespect for people, disrespect for culture, disrespect for nature.

I walk into the shop and ask the staffer behind the counter for some saddle feathers.

"Everyone wants hackles. Can't get them for love nor money. Blame Hannah Montana."

Hannah means nothing to me, but I suppose I am in Montana. "Pardon?"

It turns out that the TV character Hannah Montana (Miley Cyrus in real life) recently wore long saddle hackles as hair extensions, and instantly created an American teen-fashion craze that has bled fly shops dry. "You'd think that the breeders and suppliers would have more respect for their traditional market," the clerk laments. "But no. Mind you, they'll come back begging when the craze turns out to be short-lived, and we'll be silly enough to forgive them." He leads me to his laptop and calls up some images of Hannah's feathered hair. I wonder out loud how such a feeble look could become so trendy.

"Perhaps it's got something to do with Montana's Indian history?" says a random customer who has joined us at the computer as if he were a longtime friend. I think he means the US state, not the TV character, but by now everything seems so surreal that I can't be sure. I look again. Miley/Hannah—is there any delineation between reality and fantasy anymore? Looks sort-of pretty, I guess, but stunningly insubstantial. Miley's feathers seem, I don't know … pretentious? Disrespectful? Plain stupid? You wouldn't dare say that about Crazy Horse or Sitting Bull, that's for sure.

"You mightn't be able to buy hackles, but it sure looks like you've raised some," says the customer to the clerk, before introducing himself as Brendan. Then he turns back to the clerk and says, "How about you show our Australian friend here a bit of the Skeena steelhead episode of the *Fly Nation* TV series?"

At first I think that the Skeena angler is Hannah Montana, but it turns out that she's an American legend by the name of April Vokey.

In the back of *Trout Fishing* is a photo of Dan Bailey with a twenty-four-pound steelhead from British Columbia's Babine River, a tributary of the Skeena. The steelhead is dead, like all of the fish in Brooks's "Gallery of Trophy Trout." In the old days, any of Dan's customers who caught a trout of over four pounds was encouraged to have the fish outlined on paper and displayed along with the angler's name, date of capture, and place caught. Dan proved to be a progressive, and the practice was stopped in order to promote the new catch-and-release ethos, but the original silhouettes—over 300 of them—are still proudly displayed high on the walls of the shop, and I am invited to take a closer look.

Forget about the hackle feathers—it is enough just to be here.

SCIENCE AND FLY FISHING

Montanans, with their culturally ingrained friendliness, seem to be even more willing than other fly fishers to strike up warm conversations with complete strangers. As I peruse Dan Bailey's walls I find myself talking to two more locals who have recognized my

Aussie accent and feel compelled to welcome me to their state and country. I respond by talking about the ways in which American literature has influenced my fishing. "I first heard about this shop in Joe Brooks's *Trout Fishing* when I was eleven years old," I begin. "Brooks didn't really say much about Bailey himself, though; he seemed more interested in the various flies Bailey tied. I only recently discovered that Bailey was a scientist, that he had a master's degree in my pet interest, physics."

"Well, fly fishing is the most scientific way of fishing, isn't it? I'm Ted, by the way, and my buddy is Hugh. In a way, I think we're all scientists."

"How so?" says his buddy with a mischievous grin. One of their endearing quirks is that they constantly attempt to embarrass one another.

"Well, we follow the scientific method, don't we? We observe things and come up with ideas to explain what we've observed. We design experiments to test our ideas and see whether what we observe holds true in other circumstances. We get our educated peers to review our work and replicate and expand upon our experiments."

Hugh turns to me and winks. "Ted thinks he knows what he's talking about because he's a microbiologist. Actually, all he does all day is scoop shit out of sewage ponds."

I play devil's advocate. "Surely the thing fly fishers are worst at is abandoning old theories when the evidence fails to support those theories."

"How so?" says Ted's buddy.

"Well, for example, the idea that you need to match the hatch to catch the fish persists in spite of the fact that it's so easily refuted."

Hugh is about to argue, but Ted butts in. "Sure, everything in science is merely a theory, and even 'laws' need to be tweaked or even abandoned when they no longer conform to new facts. But scientists are only human—it's hard for anyone to abandon nice ideas, especially ones they've valued since childhood. My point is, fly fishing is at very least *akin* to science. Those who dismiss this idea out of hand usually point out that anglers' experiments are rarely robust, and that the peer-review process is unacceptably chaotic. But the one part of the equation the best fly fishers excel at is observation. This is why their insights so often beat those of the so-called professionals."

I am reminded of the park managers who couldn't see the implausibility of an unknown person transferring hundreds of adult mackinaw by road to Yellowstone Lake. Of researchers who couldn't work out that Coleridge salmon were feeding in wind lanes. Of people who do not understand that lakes are laced with currents. Of early aquaculturists who embraced hatcheries as a panacea. Of modern fisheries managers who remain addicted to hatcheries. Without a doubt, long periods of observation combined with a modicum of inquisitiveness give individual fly fishers a type of knowledge that cannot be obtained from any other source. Let's face it, no one is willing to *pay* anyone to spend thousands of hours looking for nothing in particular, and even if they did, the looking would count for nothing in the absence of passion and interaction. This, I think, is why the best fly fishers are often so "intuitive"; why they are able to see things that other experts overlook.

"Good fly fishers are good at observation," Hugh agrees, "and they are good at conservation too. Where would the conservation movement be without our advocacy and financial contributions?"

Ted, through habit I'm sure, chooses to be contrary. "Maybe we're just selfish. Now, Greg here tells us he has recently fished Yellowstone Park, and sure, there are conservation battles to be had there, but they would be fought even if there were no fly fishers. We are more interested in preserving our right to fish, which I suspect isn't quite so high on the agenda of other conservationists."

I tell my new buddies that I think it's too easy for people to say that advocacy is not much needed in places like Yellowstone, that I'm old enough to have seen many "well-protected" environments utterly trashed—even national parks. It seems to me every gain in conservation is temporary, every loss permanent; that every

Greg French

generation has to foster its own advocates. And I wonder if such dedication is really possible if people are involved with nature only at arm's length, say, by spending eleven minutes watching spawning fish from the Fishing Bridge.

Hugh sighs. "At Trout Unlimited we've been petitioning the government to do more to look after our cutthroat trout, saying that it's a matter of state pride, what with it being the state fish and all. And one of our leading politicians actually responded by saying that if the state fish was endangered, we could easily solve the problem by finding another state fish."

The three of us emphatically agree that when you fish, the natural environment is not merely a nice concept, it becomes part of your day-to-day life; that even when you are not actually fishing, you are dreaming about it; that threats to the environment begin to threaten your sense of self, your immediate well-being; that dedicated anglers *do* care enough to change the way they vote.

Ted finishes the conversation by saying, "You know, I really don't think that it's any coincidence that America's most powerful advocate for wild trout, and especially for wild native trout, was an angler before he became a scientist."

He was referring to Dr. Robert J. Behnke, of course.

Chapter 10:
Angler Biologists

GREAT ANGLER
BIOLOGISTS

We are back in Bozeman, gathered around the dining table for the evening meal. Bob and Karin have been delighting us with more stories about their time in Santa Fe—catching up with old friends, enjoying opera, visiting historic Pueblo Indian villages.

It's easy to forget that Bob and Karin are well into their seventies. Despite their encyclopedic knowledge of just about everything, from matters of the heart to all things tangible—knowledge which could be gleaned only from having lived a long life of compassion, inquiry, and travel—they are absurdly fit and young at heart. They converse enthusiastically about every topic Frances and I care to touch upon, including family, birdlife, politics, and town planning. We laugh out loud at their recent escapades cross-country skiing and hiking in Montana; and listen in wonder about their recent trip to the Serengeti, the real one in Tanzania. And we are filled with warmth at the way they gently tease each other and surreptitiously touch hands. It seems to me that they are as optimistic about their future as they were when they first met at college.

Karin gets up to serve dessert, and I ask Bob if he has ever considered living anywhere else in the United States. "Well, I could never live on the Eastern Seaboard: too busy and impersonal. I love Montana, of course, but my heart is more in the desert states. I've suggested to Karin that maybe we should move to New Mexico, but Karin, you know, she has her roots here. She says I can go live in Santa Fe if I want, but reminds me what a long commute it would be back to Bozeman whenever I wanted to see her. I figure I'm not quite ready for that."

Karin comes back into the room, grinning approval, and moves the conversation along by asking why I am so keen to interview Dr. Robert J. Behnke. I tell them how, not long after I resolved to visit Yellowstone, I heard of the passing of Dr. Bob McDowall, the giant of Antipodean native-fish research. "McDowall was one of the great angler-biologists, the last of a group who epitomized a rare combination of observational skill, unbridled passion, and clever pragmatism. I despaired that so few of my mates knew who he was. I guess I feel compelled to interview Behnke before it's too late."

Bob immediately begins quizzing me about who I would include in my list of "great angler-biologists," and I find that, in order to contextualize my choices, I have to talk about the history of trout in Australasia.

Europeans settled in Tasmania in 1804 and the colonists soon discovered that the island had no native trout or salmon—indeed, no native fish that could be deemed a sporting fish. But it did have ideal salmon habitat—large rivers that flowed into cold seas.

Serious efforts to introduce salmon commenced in 1854 when James Youl began to study ways of retarding the development of living ova so that they might survive a three- to four-month voyage across the tropics to the Antipodes. The motivation was entirely commercial—to establish a salmon industry—and in 1861, the Tasmanian Government created the Salmon Commission. By 1862, ponds and races had been prepared on the banks of the Plenty River, a tributary of the Derwent, which flowed past Tasmania's capital city, Hobart.

After several failed attempts, success came at last in 1864 when 100,000 salmon ova were dispatched from England. Youl had not intended to send brown trout, fearing they would outcompete his precious salmon, but last-minute gifts of brown trout ova from renowned naturalists Francis Francis (2,000) and Frank Buckland (1,000) ended up being included in the shipment.

Although Youl instructed that the trout be left with the Acclimatisation Society of Victoria when the ship docked in Melbourne, the shipboard superintendent, William Ramsbottom, would not give them up and sent them on to Hobart.

On April 21, 1864, live ova finally arrived at the Salmon Ponds: about 30,000 salmon ova and 300 brown trout ova had survived. The rest is history: salmon failed to establish, brown trout proliferated.

Hatcheries also proliferated, both in Tasmania and in New Zealand. This was not just because there were many virgin river and lake systems to be seeded, but also because anglers were in constant fear of the fish dying out or becoming inbred. Such fears seem quaint now, but in those days anglers struggled to understand why the fish in newly stocked waters started off growing big and fat, then after a few generations their progeny tended to be much smaller and leaner. The mechanics of population dynamics wouldn't be properly understood until the mid-1900s.

In any case, in the great majority of rivers and lakes, liberations of hatchery fish abjectly failed to improve the fishing. The first person to properly document this startling reality was New Zealand angler Derisley Hobbs. Hobbs had no formal training in biology, yet in 1948 he published one of the most influential books ever written on fisheries management—*Trout Fisheries in New Zealand: Their Development & Management*—in which he highlighted the unintended negative consequences of hatcheries. His conclusion that "it is a sound common sense rule of fisheries management to leave well alone" caught the attention of many of the world's most influential fisheries biologists and ecologists, several of whom were eager to test his ideas.

In Tasmania the Commonwealth Scientific and Industrial Research Organisation soon commissioned a study into trout population dynamics, employing an Australian-born biologist, Aubrey Nicholls, to undertake several years of intensive research. Nicholls began his work in Tasmania in 1949, and eventually concluded that the century-old practice of boosting wild stocks of trout through

189

releases of hatchery fish was not only biologically unnecessary but a waste of time and money.

Hobbs and Nicholls's work had an immediate influence in America. Biologist Paul Needham set up a research station on Sagehen Creek, part of the Lahontan catchment on California's eastern slopes of the Sierra Nevada mountain range, and one of his employees, the young Robert J. Behnke, soon proved that in healthy water, wild trout—in this case, nonnative brookies and rainbows—could easily sustain optimum population densities without the need for supplementary stocking.

This group of fisheries researchers—Hobbs, Nicholls, Needham, and Behnke—I have come to admire above all others, not only because they were the first to prioritize the preservation of habitat and gene pools over the production of hatchery-reared fish as the best way of improving fishing for all, but because they were so uncannily intuitive and *effective*.

In the late 1950s, Hobbs moved from New Zealand to Tasmania, where he was given the task of designing the structure of Tasmania's new Inland Fisheries Commission, before going on to serve as the agency's first head commissioner. During his tenure, from 1959 to 1964, he demonstrated political mastery in implementing Nicholls's recommendations, putting Tasmania at the global forefront of sustainable fisheries management, again inspiring fundamental change in the management of American fisheries.

Karin asks me about Behnke's upbringing and the things that led him to become a fisheries biologist, and I concede that I know very little. "You can easily find all of Behnke's work by surfing the Net, but there's virtually nothing about his personal life. I can't work out if he's modest or a Luddite. All I know is that in 1983 he began writing the 'About Trout' column for Trout Unlimited's *Trout* magazine, that he retired in the year 2000, aged seventy, and remains a professor emeritus of the Colorado State University."

I am looking forward to the interview, but I have my regrets too. In order to find time to do it, I've had to abandon my plans

to hike into the nearby Absaroka-Beartooth Wilderness to fish for
golden trout.

BEHNKE

Frances and I arrive at the Bozeman airport just after sunup. It may
be a rural facility, but the security checks take forever. By the time
we take off, the sky is perfectly blue. Our seats offer spectacular
views, first of Yellowstone National Park, then of the heartland east
of the Rockies, which are in the grip of the hottest drought since
at least the 1950s.

From the air, Denver reminds me of Melbourne: similar size,
same flatness; similar sprawl, same dryness. However, the airport
turns out to be a domestic air-transport hub, much bigger than I
expected. Inside, we are surprised by signs telling us that the toilets
are designated tornado shelters, and by the fact that we have to take
a train between terminals. Yet everything is logically organized, the
people are all helpful, and we soon find ourselves in our rental car
driving north toward Fort Collins.

The Coloradan landscape is relentlessly rural and parched. I
amuse myself by watching bald eagles soaring on thermals, and
relive the impact Rachel Carson's *Silent Spring* had on my formative years. I wonder yet again if we will ever find the strength to
respond to the threat of climate change in the same decisive way
we responded to the threat of DDT.

Fort Collins is a delight: a university town of less than 150,000
people where we find the first genuinely good coffee we've had in
America, and real loose-leaf tea too. Being landlocked, but with
an inescapable air of university high jinks, it seems natural that
a beach theme pervades: surf shops, cars with registration plates
like SALTH2O, even a Bondi Beach Bar. I love the street theater,
and the fact that the shops all have water bowls on the sidewalk for
patrons' dogs. I am about to pinch myself to make sure that I'm not

dreaming about this town being in America when we are pleasantly accosted by a group of earnest parents and grandparents asking if we could possibly spare a dollar or two to help supply dictionaries for local primary schools.

In the early afternoon we drive to the eastern edge of town, to the address that Behnke has given us. We pull up outside a modest clapboard home, and a woman, presumably Behnke's wife, Sally, comes out to meet us. We say hello as we walk toward her, and she says hello back, but her tone sounds defensive—as if she is mildly resentful of our arrival. Perhaps her husband is being constantly pestered by journalists, fisheries biologists, and anglers, and she has become protective.

Suddenly a man appears at the woman's side. "Greg and Frances? I'm Bob and this is my wife, Sally." He is older than Karin's Bob, and less physically fit, but the brightness in his eyes reveals a razor-sharp mind. His tone is more encouraging than Sally's, but it contains enough reserve to keep me ill at ease. Frances and I make eye contact. I'm not a natural at interviewing, and wonder if I'm out of my depth.

"It's drier than I thought it would be," says Frances.

"Sad, isn't it?" says Sally. Soon she is leading us on a tour of the shriveling remains of their once-wonderful garden. Their heartbreak is the same that many of us from southeastern Australia have already endured. I came to America wanting to know why mackinaw and whirling disease had suddenly boomed in Yellowstone Lake, and the answer has probably been staring me in the face all along.

While Frances and Sally examine the plants, I natter with Bob. He tells me that he was born in 1929, that he married Sally in 1963, and that they ended up with two children: a son who became an electrical engineer, and a daughter who became a bassoonist. Then he invites me into the lounge room. Frances stays with Sally. Soon we two men are sitting opposite each other in leather armchairs, and much more relaxed now, I finally get to ask the world's leading authority on salmonids what first prompted his interest in trout.

"My earliest memories are of fishing for yellow perch in a mill-pond in Stamford, Connecticut. Dams and pollution had destroyed our local populations of native brook trout before I was born, but each spring the nearby Rippowam River was stocked with hatchery-reared brook trout. The day I caught one on a worm, I ran home, placed it in a pan of water, and watched it swim. I thought it the most beautiful thing I had ever seen, and was hooked for life. At that time I did not know, and would not have cared, that it was a hatchery fish. My transformation to an advocate for wild native trout was a gradual process based on learning and experience."

I notice that, when on topic, Bob speaks as if he's lecturing, and I recognize that much of what he says has been lifted almost word-for-word from his books. I am guilty of this myself: when you have considered an argument in great depth and have had the opportunity to refine your ideas in writing, the final edit does tend to roll off the tongue.

"In the 1950s, I was drafted into the army, and I went on to serve in Japan, then Korea. Along with a rifle, I carried a fly rod. I explored mountain streams and found yamame and iwana. What, I wondered, was their relationship to the American trouts? Upon discharge, I was surprised to learn that the government would pay for me to attend college and study fish. In 1957, I completed my honors degree, then I was contacted by Paul Needham from the University of California and invited to work as a graduate research assistant on Sagehen Creek. We were checking to see if wild trout could sustain quality fishing without the need for hatchery fish, mirroring work done a year or two beforehand in your home state of Tasmania by Aubrey Nicholls."

I feel proud that Bob is talking about, and clearly respects, Nicholls.

He says, "Of course it must be remembered that Nicholls's work may well have been ignored but for the efforts of Derisley Hobbs."

He's talking about Hobbs too? I ask Bob to nominate a list of the most influential fisheries scientists in the world and he doesn't pause to think: Nicholls, Hobbs, McDowall, and Ricker.

Greg French

Well, how about that: three out of the four are from my remote corner of the world! I feel awkward in having to admit that I don't know much about Ricker.

"William E. Ricker is an American who in 1959 wrote an unpublished manuscript about how discrete populations of sea-running salmonids in the same river adapted to local conditions by various minor behavioral traits. He assumed these to be genetically ingrained. I read one of about one hundred copies for comment, and was greatly impressed."

"What did you do after that?"

"Then I became interested in the long-gone *native* trout of Sagehen Creek: the legendary Lahontan cutthroat trout. By the 1950s this unique lineage of cutthroats was thought to be extinct in the wild in its native range in the Lahontan basin. However, I felt that there must be some remnant populations above barrier falls on headwater creeks, and around 1960 I hiked into the East Carson River and found native trout in two headwater tributaries. After this I became a passionate advocate for the preservation of what remained of America's native fish."

I ask if he has had any word on the Pilot Peak fish released into Pyramid Lake in 2006.

"Well, we're still waiting." Somehow his facial expression and body language indicate a sly confidence that something remarkable will soon be reported. As an aside, he also mentions that Lahontan cutthroat ova were transported from the original Pyramid Lake hatchery to the California Acclimatization Society from where, in 1875, a batch eventually found its way to New Zealand. "Unfortunately, nobody knows what happened after that."

I ask if he has any theories on how mackinaw came to be in Yellowstone Lake and why they have not greatly impacted the cutthroat in Heart Lake. He replies earnestly, "It is inconceivable that large numbers of adult mackinaw were transferred to Yellowstone Lake from Lewis Lake without detection. A more likely explanation is that they were liberated during the hatchery era. The hatchery staff used to hold open days when the public was invited to inspect hatchery operations. Often, most or all local salmonid species were displayed in aquaria. I doubt that these display fish were always killed after the event. As for Heart Lake, it contains a wealth of fish other than cutthroat on which the mackinaw can feed, principally Utah chub, and these probably mitigate against the worst predation scenario."

I am also curious about why many subspecies of American trouts, especially cutthroats, are identified by their coloration and patterning. After all, the brown trout in Tasmania and New Zealand belong to a single subspecies (*Salmo trutta trutta*), but display many color markings, and these colors seem to be determined by environmental factors alone—sea-run fish are silver, those living in small creeks are often dark brown with red spots, those in Tasmania's Western Lakes are usually gold with heavily spotted heads and tails.

"The problem with brown trout in your homeland is the same as it is in America. I consider these transplanted fish to be an all-purpose or generic type of trout resulting from the mixing of several races from multiple locations, all with different life-history adaptations. Natural selection processes operating for more than a century in North America—and Tasmania and New Zealand—will have fine-tuned brown trout to their new environments. Given the genetic diversity of the founding stock, it is hardly surprising that a range of color variations and life-history forms have arisen. Would vibrant runs of sea trout have arisen from a pure strain of ancient landlocked stock? Most likely not."

Bob also says that if I want to know what the various pure-strain brown trout look like, and how their life histories vary from other brown trout, I should read James Prosek's book, *Trout of the World*. But he laments, "Sadly, many of these pure-strain fish are now extinct, having been hybridized out of existence through interbreeding with hatchery stock, most of which were derived from parent stock of Atlantic origin."

Greg French

Our conversation moves on to how to define *wild trout*, and on this subject Bob is definite: "The word *wild* comes from *willed*, meaning recalcitrant. The word *domestic* means compliant. A wild fish is one not raised in a hatchery, one that is the result of natural spawning and natural selection. Natural selection cannot occur in a hatchery. And artificial selection occurs even if hatchery managers are not actively selecting fish for any trait."

So is putting hatchery fish into wild fisheries a problem only for native trout?

"It depends on what you are trying to preserve. If you are trying to maintain quality fishing for nonnative fish, the answer is complex. Wild trout, even nonnative wild trout, have a very real intrinsic value, an intangible aesthetic value best extolled by Roderick Haig-Brown. But it is hard to quantify. In many American states, the continuing emphasis on stocking 'catchable' trout results in a small proportion of anglers being heavily subsidized by all other anglers. The value to the economy of an angler spending a day fishing for wild trout is always higher than a day spent fishing for hatchery-reared catchables. Stocking is best done in small ponds, typically less than 100 acres, with good access. Here the percentage return is much, much higher than for fish released into streams or large lakes."

We agree that wild fisheries are vibrant, stable, and affordable—and domestic fisheries are bland, unstable, and expensive—but Bob stresses that there is more to the story. "With brown trout in your country, a century of evolutionary fine-tuning will have resulted in perfectly adapted wild fish. It would pay to leave well alone."

I wonder if releasing hatchery fish into wild fisheries can ever be justified.

"Not in robust wild fisheries. But as a last-ditch effort, yes. Hatcheries have been critical in bringing back near-extinct species such as the Lahontan, Gila, and Apache trouts."

Frances and Sally return from the garden, and Sally insists that we retire downstairs for "a little libation."

197

Greg French

Bob smiles conspiratorially. "We find that a glass of whiskey keeps one young."

We are all completely at ease with one another by now, and Bob and Sally enthrall us with their stories of working behind Russia's Iron Curtain in the 1960s. Then we exchange tales about Ireland, Slovenia, and Mongolia. Bob is happy to explain all the latest genetic findings, and updates me on the status of species, subspecies, and races of many of the world's salmonids. The hours fly by.

Just before we get up to leave, I mention to Bob that many of the men he has lauded had no formal training in science. Johannes Schöffmann from Austria (the preeminent authority on European trout) was a baker. Kent Andersson from Sweden was a writer of popular articles in local papers. James Prosek from America is an artist. Even Derisley Hobbs was an amateur. Furthermore, many of those who did have a university background were keen anglers anyway, including Phil Pister, the American fish biologist at UC Berkeley who became internationally renowned for stressing that predators and competitors were "good" species because they helped make up the complexity of nature, and Bob Smith, the author of *Native Trout of North America* (1984). I also cite Samuel Snyder, an environment educator from Alaska who wrote to *FlyLife* stressing that the main problem in the conservation of American trout and salmon was the lack of youth anglers, "because they will be tomorrow's advocates." I suggest that anglers seem to make the best fisheries biologists.

Bob's answer comes as a mild surprise: "Clever anglers spend a lot of time observing, and in science, clever observation is invaluable. There is no doubt that amateurs with a love and enthusiasm for the subject can discover new information and make a significant contribution to a better understanding of trout. But some of my best students have not been anglers. Women, I find, often come to the discipline from a different background, and because of this they frequently offer refreshing new perspectives. One of the problems with letting anglers set the agenda and make the rules is that they are often self-serving. Science should not pander to the hook-and-bullet brigade."

One last question: What does he consider to be his most important contribution to fisheries science?

"My students."

I don't for one moment doubt his sincerity, especially not after he has described many of his students' astonishing achievements, but to me Bob Behnke is mainly to be lauded because, more than anyone on the planet, he has fostered an ongoing respect in the angling community for science and evidence-based management.

Greg French

Chapter 11: Comparisons with Mongolia

LOWER YELLOWSTONE RIVER

We have returned to Bozeman, and Bob and Karin decide we should go for a float. They want to know if I have a preference for any particular river, and without hesitation I nominate the middle Yellowstone near Livingston, mainly because I want to see for myself what has happened to the native cutthroat.

For many decades after 1915—when the Wind, Bighorn, Rosebud, and Tongue Rivers are presumed to have lost their native Yellowstone cutthroats—the species fared reasonably well farther upstream in the vicinity of Livingston and Gardiner. But after almost 100 years of coexistence between rainbows and Yellowstone cutthroats, it appears that the barriers to hybridization have broken down. Behnke fished the Yellowstone River at the turn of the millennium and noted that near Livingston, most trout were rainbows, and hybrids seemed more rainbow than cutthroat. At Gardiner, too, a lot of hybridization occurred, even if most hybrids seemed more cutthroat than rainbow.

Bob and Karin take us to the River's Edge shop, where Frances and I buy Montana fishing licenses and meet our guides. Chester is going to handle Bob and Karin's boat, and Zach is looking after Frances and me. We'll be fishing from hard-sided McKenzie drift boats.

We spend half an hour or so driving to our destination before putting in at Carter's Bridge, three miles upstream (south) of Livingston. "We're going to do the Town Stretch, a drift of about eight miles through Livingston to the Highway 89 Bridge."

While Zach and Chester ready the boats, I notice a delightfully simple Fishing Access sign incorporating little other than a brown silhouette of a fish and the name of the access point. Another sign, authorized by Montana Fish, Wildlife & Parks, proudly broadcasts, "This site has been adopted by and cared for by Montana Whitewater."

In Montana, the importance of access rights for anglers and other members of the public has long been acknowledged and promoted. Although there have been legal challenges, the order of the day has generally been cooperation between government agencies, anglers, canoeists, and private landowners. In part, this is because state law enshrines the right of the public to use rivers for recreational purposes up to "the ordinary high-water level." So if an angler can gain access to the water, he can drift through private land or walk along the banks without fear of eviction. Furthermore, state trespass laws allow public access across private land unless the landowner has posted notices specifically excluding access at each perimeter gate and wherever a stream enters or exits the property. As ranches are generally large, and public right-of-way exists along the riverbanks, negotiating access points with landowners has been a relatively simple matter.

New Zealand fishing has been more heavily influenced by American than English or European culture: the statutory authority managing the fishery is called Fish & Game (as are similar authorities in many American states), the most popular generic dry fly is the Royal Wulff, and most of the country's current management strategies were first put on trial in America (catch and release, ballot systems on the Greenstone River, etc.). Kiwis even pronounce "cicada" with a hard "a," the same way Americans pronounce the "a" in tomato, and they are increasingly likely to boast of catching "a real hog of a fish." Pronounced *haarg*, of course.

Greg French

One of the better things the Kiwis have mimicked is Montana's Angler Access program. This works well in New Zealand, mainly because of the Queen's Chain regulation, a legal curiosity dating back to the mid-1800s that gives the public the right of access to a strip of land twenty-two yards wide along both sides of any stream, and alongside any lakeshore or coastline. Certain aspects of the Queen's Chain have not been tested in law, but this is irrelevant; the idea is socially ingrained and no one seems likely to take a broad-scale appeal to the High Court anytime soon. Rural properties in New Zealand—stations—tend to be big, so in order to formalize access across private land, statutory authorities have had to negotiate with relatively few landowners. Again it has been a rather simple matter to negotiate agreed access points and routes.

Tasmania borrowed the Angler Access idea directly from New Zealand, but without ever understanding the history or logistics underpinning the concept, and the results have been contentious, mainly because the state has no enshrined rights of access along privately owned waterways, and rural properties tend to be relatively small. Prior to rolling out the program, I had never been asked to leave a property. Since the program began, "No Fishing" signs have sprung up like weeds.

Woody Guthrie would have been disgusted, I'm sure. After all, he believed that the land is *our* land:

> As I was walking I saw a sign there
> And that sign said "No Trespassing."
> But on the other side it didn't say nothing.
> Now that side was made for you and me.

Zach mentions that the fishing has been tough of late and that rises are unlikely. I can tell that he is seeking permission to *not* use dry flies. I ask if he can show us exactly how he himself would fish under these conditions, and to talk us through what he is doing. Accordingly, he rigs us up with large weighted nymphs and bizarre balloon indicators, and once on the water, in his charmingly calm and understated manner, he continually points out the right currents and shows us how to perform clever little casts that are easy to mend.

Frances hooks the first fish—a large mountain whitefish (*Prosopium williamsoni*) weighing more than two pounds—and we are ecstatic.

"I'm glad you're happy. I like whitefish too. But, you know, most people hereabouts don't much care for them. Everyone seems disappointed if it's not a trout."

"Native salmonids in wild habitat—it doesn't come much better than that for us," I say.

"Well, I wish everyone was like you. When whitefish are plentiful, people take them for granted, even think of them as a trash fish. But when whitefish disappear, folks suddenly realize what a godsend they were."

Disappear?

"The whitefish are still doing pretty well here in the middle Yellowstone, but elsewhere it's a different story. About five or six years ago, guides fishing the Madison noticed a drop in the number of whitefish they were catching, and it's gotten worse every year since then. Hardly any are being caught these days. The population is said to have fallen from maybe 15,000 whitefish per mile to next to zero. Reports have since come in from all over Montana and Colorado that whitefish numbers have crashed. It's a sobering turn of events. Whitefish are a sign of a river's overall health. Trout eat eggs and juveniles; adults are important food for ospreys, eagles, and otters. Fisheries biologists are pretty worried."

"What's caused the problem?"

"Most people think it might be whirling disease."

God, I think, *not a fourth hatchery-induced disaster on the long-suffering Madison.*

"If people are honest, though," Zach continues, "no one really knows what's going on. It could be habitat loss or climate change.

Greg French

More likely it's a combination of lots of things. There's no guarantee that the Yellowstone won't be next."

Frances catches half a dozen more whitefish, and I catch two, and I'm even more grateful now than when we caught the first fish.

Frances catches a one-pound rainbow, then a two-pound rainbow, then a two-pound brown. Now I'm not so much grateful as jealous.

An osprey snatches a large rainbow from the water just ten yards ahead of the boat. It alights in a nearby tree and starts picking away at its catch. We drift on a little farther, and enjoy a fine lunch on a gravel bank with Bob, Karin, and Chester. Then we head off downstream again.

By the time I see the Highway 89 Bridge, Frances and I have boated a dozen rainbows, half a dozen browns, and more than two dozen whitefish. The only minor disappointment is that we haven't caught an obvious cutbow hybrid. But with just a handful of casts left to go, even that situation is rectified. We stare at the chubby fish glistening metallic pink against the translucent silicone mesh of the landing net. The only feature that differentiates it from the rainbows we have caught is the large and brilliant red slash under each jaw. Zach reminds us that some or even all the rainbows we have caught and released today probably carry some cutthroat genes. But this fish is special to me because it is the first time I feel that the native trout of the middle Yellowstone have not yet been completely obliterated.

After such a great day, it occurs to me that when I am overseas, I hire guides only when I need them to supply some crucial ingredient other than the actual guiding, in this case the drift boat itself. Yet the real value of today's experience, as always, has been the conversation with the guide and the chance to learn about local lore and techniques.

At home in Tasmania—and especially in New Zealand—guiding is a vexed issue, but it seems to me that all the common arguments used against it don't stand up to rational scrutiny.

Surely the idea that publicly owned fisheries should be mostly managed for locals, not guided clients, is the same as suggesting that anglers should mostly confine themselves to waters close to their place of birth or residence. For proponents of this view, travel and the desire to broaden one's horizons are to be discouraged.

The lament that "Guides are popularizing my favorite places"? Well, perhaps we would all do well to think about how we got to know about our favorite places in the first place. Did someone tell us about them? Did we read about them? Find them on a map? Our information resources are diverse—we rarely find out about anything entirely on our own. And anyway, I have learned firsthand that popularization is important. A proposal in the 1980s to privatize Tasmania's Little Pine Lagoon was scrapped when a majority of the 1,700 anglers who fished it, including the guides, lobbied the state government. The private land around the lagoon was subsequently acquired by the state in order to create a special fishing reserve. This would not have happened if there were a mere handful of lobbyists.

As for the argument that "Guides are overfishing our waters," it's my experience that guides don't fish, their clients do. And exactly why a guided visitor should be considered less tolerable than an unguided one is beyond me.

The difficulty of regulating visitor numbers for social reasons is that the definition of "acceptable use" changes with time. As already mentioned, if you ask a population of anglers (or trampers, or skiers, or surfers, for that matter) what they want, the great majority say that they want visitation to stay "the same as it is now." But with the world's population increasing, I suspect that we'll just have to get used to sharing water. The good news is that for the foreseeable future, the fishing in Tasmania, New Zealand, and Yellowstone is completely sustainable.

In the end, I suspect that it's not really the guiding that anglers are afraid of. After all, if you read magazines—even ones like New Zealand's *Fish&Game* that constantly portray guiding as something sinister—you are essentially being guided. And if you refuse to read,

Greg French

you are simply bigoted. I think that what anglers are really afraid of is having to share water with other people. The problem is, if we are to protect our fishing from the anti-fishing lobby and environmental destruction, we will need all the political clout we can muster. We need to encourage more anglers, not marginalize them.

It is worth remembering, too, that the impression of heavy fishing pressure is often different from the reality. There were lots of boat trailers where we put in at Carter's Bridge, but we ended up having the water to ourselves.

MARK JOHNSTAD

I am going to use my last full day in Bozeman to see a river guide: Mark Johnstad, the Montanan who set up Mongolia River Outfitters (MRO). In less than two months' time I will be taking a group of six mates to Mongolia, and although the trip will be "self-guided," MRO will be supplying much of our equipment and taking care of logistics. The reason that a trip of this nature is possible is that MRO has recently upgraded to hard-sided McKenzie boats, which means that the old inflatables will be sitting idle. I will also be able to use MRO's existing ger camps, along with their drivers and camp staff. I took the "cheap" option not because it was cheap, but mainly because my mates and I all love the adventure of unguided fishing.

Bob and Karin offer to drive with Frances and me to Mark's home, which is just a couple of miles from theirs. We are met at the door by Mark's wife, Megan, who is run off her feet because the whole family is supposed to be on a plane to Zambia. Alas, their son Cooper, age six, has somehow allowed his passport to expire. Naughty boy.

Megan is a biologist, and her current work in Africa is focused on cheetahs. I reach down to pat a golden retriever and a Belgian malinois who have bounded through the house to nuzzle my legs.

"That's Chobe and Peppin. Peppin is going with us to Africa. He's a scat dog, trained to sniff out cheetah poo."

It turns out that Megan is part of a Montana organization called Working Dogs for Conservation. "One of our organization's dogs is going to be trained to sniff out nonnative species that threaten trout populations in the Rocky Mountains. The first target will be brook trout. Dogs can actually scent the fish in the water."

Wow, where does that leave all those wonderful stories from my childhood about escaped convicts eluding the posse by wading up streams and rivers? I admit to being mildly disappointed.

Megan tells us that we can find "the boys" out by the creek, and points the way down an embankment along an unformed and overgrown footpath. It doesn't take long to find them. Mark is coaching his son on the intricacies of dry fly fishing. The stream is narrow and overgrown, but riffly and full of small trout. Cooper hooks a brookie and brings it over for us to see. "Hello, Bob and Karin," he says warmly, before turning his attention to Frances and me. "You must be the Australians. I'm Cooper. Have you known my dad for long?"

Cooper removes a pocket knife from a leather case attached to his belt and says, "I hope you don't mind, but I'm going to kill this fish. It's a nonnative, and Dad lets me eat nonnatives. The cutthroats are native, and I'm not allowed to kill those."

Mark mentions that he has to go and check on what progress the authorities are making with Cooper's passport. He suggests we all come back with him for morning tea. Bob, Karin, and Frances turn to leave, but Cooper says to me, "You can stay here with me if you like. I'll show you how to catch fish." I look at Mark, who nods.

Cooper offers me his hand and says, "Can you steady me across this tailout—it's a bit deep and slippery to manage by myself."

Cooper ends up catching two more brook trout and a rainbow and then, despite my protestations, insists that I have a go. I am astounded by how articulate and knowledgeable he is. How sweet, trusting, and generous.

Greg French

Finally we extract ourselves from the creek and make our way back to the house. Bob and Karin have already left, but Frances is sipping tea with Mark while Megan remains busy getting her research equipment ready for Zambia.

Cooper offers to cook his brook trout for us, and while he sets to work in the kitchen, I talk about the logistics of our upcoming trip and quiz Mark about how he came to set up MRO.

It turns out that Mark was raised in Montana and always loved the outdoors, and when he saw the wild being sucked out of the West, he resolved to become an environmental lawyer. But before getting stuck in a bland, conventional lifestyle, he wanted to do something exciting. He had read books about Mongolia—all written in the 1920s because there was nothing available from the intervening period of Russian rule—and decided that Mongolia was a lot like Montana had been. Since he had always fantasized about going back in time to pre-colonial Montana, he contacted the Mongolian ambassador to the United States via the Washington embassy. They ended up talking mainly about conservation, especially the future of public lands in Mongolia following the transition from the centrally managed system that existed prior to 1991 to the inevitable free-market system. Siberia had already overexploited its stocks of big game and sport fish, and with a mining boom on the horizon, Mongolia seemed ripe for corruption and bribery. The question was, what could be done to help preserve Mongolia's fauna before the rot set in? They decided that the fledgling democracy might be able to learn from the successes and failures experienced during the evolution of the United States' system of parks and reserves.

In the summer of 1991, with the support of the Mongolian government, Mark traveled to remote areas all over Mongolia, helping to draw boundaries for new protected areas.

However, despite all the conservation efforts up until the mid-1990s, Mongolia's fish and game continued to disappear at a rapid rate. "A century ago, taimen were found all over Asia and much of eastern Europe, but by the time I arrived in Mongolia, their range had been greatly reduced. In Russia, robust wild populations had been relegated to the remotest corners of the Siberian wilderness, and the European variety was being largely maintained by artificial stocking. We were really worried about what the future held for Mongolia," Mark explained.

In 1994–95, the Vermillion brothers from Sweetwater Travel contacted Mark to talk about the viability of setting up a guided taimen fishing operation. Mark immediately saw the possibility of using fishing tourism to promote fishery conservation, so he assisted with logistics and helped secure the appropriate permits on the Egiyn and Üür Rivers (in the Selenge mörön drainage).

In 1997, Mongolian tourism officials approached Mark to talk more about guided fishing, and the following winter—tired of doing conservation only at arm's length—Mark began to work with the owner of Nomadic Journeys to see if guided fishing could be expanded as a tool for conservation. He decided to do a trial run on the Onon because this was the most pristine taimen and lenok habitat in all of Mongolia.

I ask Mark to list his greatest achievements. He pauses for a second and replies, "I've been lucky. Lots of great adventures. Lots of fulfilling activities. But no one ever achieves anything by themselves. We all have help somewhere along the line. For me, working with a great group of folks to help establish the world's first taimen sanctuary at the Onon River is definitely a high point. How often do you get to show that fly fishing can help change the world?"

MONGOLIA REVISITED

Rural Montana has changed dramatically over the last 150 years, whereas rural Mongolia has barely changed at all. Now, though, everything is on a precipice. Can the Mongolian government really

learn from Montana's experience in fisheries management, or are people everywhere doomed to perpetuate the same mistakes?

I intend to find out.

It is early September 2012 and our group, freshly arrived at the Chinggis Khaan International Airport, has spotted our translator, Bayanaa. It wasn't hard for him to find us; everyone else from the plane is long gone. He introduces himself and says gently that the Onon is in flood, that the fishing has been tough, and that it will continue to be tough. We tell him that Anton's gear—rods and all—was lost in transit.

Next morning, we hang around longer than planned in Ulaanbaatar (UB), but Anton's gear fails to turn up. Early afternoon rolls around before we finally get into our chauffeur-driven four-wheel-drives. Our driver's name is Bold (meaning *steel*), which we mishear as Bolt. The younger driver of the other vehicle is also called Bold. When we finally resolve the confusion about who is who, we come to differentiate them as Old Bolt and Young Bolt.

The traffic in UB is much more chaotic than it was two years ago, and it takes our drivers two hours to negotiate the first few miles out of the city center. The next thirty miles of paved road, leading to the 130-foot-high Chinggis Khaan memorial at Tsonjin Boldog, is a crumbling mosaic of potholes and takes another two hours, though I guess the trip would have been quicker if we hadn't stopped to buy drinks from roadside peddlers, or to have our photographs taken alongside camels, eagles, and vultures, all tethered and sad. Chinggis, at least, looks content. Perhaps it is because he is facing east toward his birthplace, toward our fishing destination.

The paving on the next ninety miles of "highway" to Jargalthaan is less eroded, and occasionally we hit fifty-five miles an hour. Then we turn northeast onto an "unformed road," which proves to be nothing more than a lattice of wheel ruts across unimaginable expanses of open steppe, though the dry granite soils permit hair-raising speeds of up to eighty miles an hour. Old Bolt and Young Bolt are renamed Thunder Bolt and Under Bolt.

Greg French

Most of the people we pass on the great open grasslands are on horseback, and there are no fences. The land really is how I imagine Montana must have been before the railways were laid in the mid-1800s.

Darkness falls quickly at Ömnödelger, the last tiny town en route to the Onon. We still have 110 miles to go, and our pace slows again. Periodically we meet an oncoming motorbike or hay-laden lorry, often driving without lights under the glacial stars, or a goat-laden wagon pulled by a yak or cow. Just as mesmerizing are the moments when our headlights startle a jerboa, marmot, or steppe fox. When our drivers stop to rest and smoke, and we take the opportunity to piss and stretch, the howling of wolves crystallizes in the infinite stillness.

We arrive at camp at two in the morning, and immediately walk with our headlamps to the riverbank. High and dirty: an autumn flood. It has happened only once before in the last seventeen years.

"By the tide lines, it's dropping quick," I offer. "Being able to see a foot into the water is better than nothing. Fishing will be hard tomorrow, but things will improve day by day." I believe what I am saying, but I also feel obliged to foster good spirits. We've been fantasizing about this trip for more than a year, and getting to Mongolia hasn't been cheap. I really wanted to show them what the Yellowstone River used to be like in Custer's time.

We soon discover that the camp staff have long been asleep, that our meals are cold, that the water in thermoses is tepid. None of that is anyone's fault, and it's harder on the soul when there's no one to blame. At least we have plenty of sleeping equipment, clothes, and fishing gear to share with Anton.

I initiated this trip, so I feel responsible and don't sleep well.

I am the only experienced rafter. Is there going to be a safe stretch where I can coach the rest of the team? Learning to handle the rafts at normal river flows would be simple, at least for competent people like my mates, who are all fit and have had lots of experience with dinghies. But they still need to practice the basic maneuvers.

In the morning, Peter from MRO arrives to brief us.

"How's it been?" I ask.

"Tough. Yesterday the anglers in my raft rose three taimen and landed a smallish one, so there's hope."

"Lenok?"

"You'll catch some."

Some? I was hoping that my mates would catch hundreds.

I ask Peter to mark our six camps on the 1:250,000 contour map I bought in UB. He looks hard and says that the detail is too vague for him to be certain of where everything is. He calls to a young man standing nearby. "This is Zolboo, a local from Dadal," Peter says. "Perhaps he can help us."

Zolboo glances at the map, and instantly points out the exact location of all six sites. I am certain that he is not used to looking at maps, but equally certain that he is able to instantly visualize contours as three-dimensional images.

"Zolboo is going to row one of your rafts for you," Peter says.

A guide wasn't part of our deal. I suspect Zolboo is being offered because of the flood, and I am grateful.

Zolboo is nineteen, studying fisheries management online, and working in his spare time as a river guide. He rafted the Onon last week, during the peak of the flood, and knows the exact location of the tricky bits. He assures me that we will have everyone trained in no time.

Zolboo has been learning English for only six months, but his use of the language is perfectly functional, and he is already having fun with wordplay.

Before we set off, I remind everyone that the use of barbless hooks is mandatory.

Anton stands upright, as if on a soapbox, and gives a prim and convincing recital of advice I may or may not have written. "The empirical evidence shows that barbed treble hooks on lures cause no more mortality than barbless hooks on flies. Indeed there may be a slight statistical tendency for barbless hooks to result in slightly more mortality than barbed ones, perhaps because they tend to penetrate deeper, though in all studies this tendency is very slight and could be accounted for by the statistical margin of error." Everyone giggles.

I concede that I have become frustrated with the way fisheries managers all over the world seem prone to prioritize the implementation of discredited regulations while ignoring the things that really matter. Here in Mongolia, however, there is an altogether practical reason for using barbless hooks. "Okay, guys, laugh at my expense if you like, but the last time I was here everyone was hooking dozens of small lenok. They have really small mouths, and whenever we forgot to crimp the barbs on our hooks, it was one hell of a job to remove the hook. Here, using barbless hooks really will reduce fish mortality, and also save you time."

Zolboo listens to our conversation with amusement, and then lets us know that fisheries management is a matter very close to his heart. Tomorrow he and Yuruult, the local guide I met on my first trip here two years ago, are going to chair a meeting about taimen conservation in Bayan Adarga. High on the agenda is an offer by French and Czech interests—championed by a politician's wife—to establish a fish hatchery on the upper Onon in which taimen would be produced to supplement stocks in a number of Mongolian rivers.

So here we go again—can the hatchery hydra *ever* be slain? This river, the Onon, is exactly what the Yellowstone used to be like before it was irreversibly degraded by hatchery operations. It is exactly what everybody wishes America's trout fisheries were like today. Things cannot be made better than they already are.

I mention that I have recently been talking to Behnke, and stress that he strongly advises against hatcheries. "The Onon is pristine and already carries precisely the number of taimen and lenok that it can support. The carrying capacity is dictated by the number of holding pools, the quantity of available food, and other physical and biological restraints. Putting more fish in the river won't increase the carrying capacity; the stockies will either die or compete with wild fish for food and space. The results won't be pretty," I tell Zolboo.

I also mention the problem of genetic contamination, pointing out that Amur trout and lenok are so closely related that there is scarcely any genetic difference between them, that the behavioral barriers that prevent hybridization could be lost if hatchery fish were introduced to the Onon, just as has happened with rainbows and cutthroats in many North American waters.

Taimen and lenok are also extremely closely related, so much so that it is said they occasionally produce fertile hybrids in the wild. The introduction of hatchery fish could cause these two species to morph into a single hybrid population, resulting in fish that are virtually indistinguishable from typical lenok.

Hatchery fish are also likely to result in more subtle but equally problematic genetic pollution between fish of the same species, leading to a loss of diversity similar to that which has come to plague North America's coastal rainbow trout.

"And, of course," I conclude, "these genetic problems are additional to the usual problems of hatcheries exacerbating disease and having a negative impact on tourism. Really, who is going to spend tens of thousands of dollars traveling to Mongolia to catch disease-ridden hatchery stockies?"

Zolboo says that initially many locals thought that a hatchery would increase the number of taimen, so they supported the idea. But Yuruult, other locals, and himself are well read and have long understood that the arguments in favor of hatcheries are often simplistic and wrongheaded. He assures me that he and his colleagues have been widely informing local communities of the damage hatcheries have inflicted upon wild fisheries throughout the world.

Our first day fishing is tough—the lenok refuse to rise in the fast, dirty water and even nymphing proves next to useless—so we end up having to fish for taimen the whole time. Miraculously, we land a giant taimen or two, and also some Amur pike and whitefish (*Coregonus chadary*). We are upbeat at dinnertime, even if Zolboo has gone away for his meeting and we miss his company.

Greg French

When Zolboo arrives back at camp the next morning, he is relieved to tell us that, having reviewed the evidence, all soums (districts similar to municipal councils) in the Khentii aimag (province) are now firmly against hatcheries, and they will be presenting a united voice to the Mongolian government requesting that any proposals for hatcheries on the Onon be abandoned.

I am as relieved as Zolboo. But surprised. When hatcheries were first discussed in other parts of the world, most local anglers did not want to look at the evidence against them, much less believe it. Many anglers in Australia, New Zealand, Europe, and America still do not want to look at the evidence, or believe it.

Historically, fishing has not been a popular pastime in Mongolia. Taimen have a mythical quality that has inspired reverence, but the other fish have largely been ignored. Bayanaa tells me that, in part, this is because the local traditions of Shamanism and Buddhism shun the consumption of cold-blooded animals.

I suppose that since both Shaman and Buddhist followers revere nature and understand the importance of leaving well enough alone, the Mongolians have been especially well disposed to reject the idea of hatcheries as a panacea. But there is no doubt that the presence of MRO has fueled local interest in fishing, and helped foster politically powerful and environmentally aware fishing clubs in several villages along the river. In any case the locals, having developed a strong affinity with the fish, now seem determined to protect what they have come to love. Participation really does equal advocacy.

The fishing remains tough over the next few days, though everyone lands at least one big taimen and a few larger-than-normal lenok. By now we have learned to look for and target fish that are feeding on schools of baitfish. We have also abandoned our eight-weight rods and big flies in favor of equipment that's easier on the wrists and elbows.

The last day is again cold, gray, and tough. By noon a couple of large piscivorous lenok have been landed, but no taimen. We pull over on a gravel bar for lunch, just downstream of the Dadal bridge.

After a time, Zolboo picks up one of our five-weight rods and studies the tiny, black wet-fly. "Do you mind if I try?" he says, looking at the shallows where a small anabranch separates from the main current.

It takes five casts, the only five casts that Zolboo has made all trip. The fish runs upstream, then back downstream into the main current. It dogs heavily along the seam, and then around and down in the main hole. There is frantic talk of launching a raft lest it makes off down the river proper. But Zolboo handles the animal like a whisperer handles a horse, and after fifteen minutes the great taimen is netted in the shallows at our feet. It is the biggest taimen that Zolboo has ever caught on a fly. There are hugs all round, and photographs and congratulations.

The only sad thing for me about this trip is the inability to catch big numbers of lenok. Oh well, I guess even Custer's troops must have had hard days during major floods.

Back in our hotel in UB, I Google the Onon hatchery issue, and discover several papers that Yuruult has co-authored with his friend Gaana and Dan Bailey from Montana. Dan Bailey from Montana? I wonder if he is any relation to the original owner of the fly shop in Livingston.

In any case it seems to me that Montana can be proud of its influence in Mongolia. Perhaps the tragic history of the Yellowstone cutthroat trout can help prevent similar heartbreak from occurring elsewhere in the world.

Greg French

Chapter 12: The Tasmanian Connection

BACK HOME IN TASMANIA

It's April 2013 and I'm home in Tasmania catching up with a back-log of writing assignments, struggling to stay focused. The fate of the Yellowstone cutthroat still bothers me, and events at home and abroad seem likely to provide further insights into the fish's long-term chances of survival. There's so much I want to follow up.

I log on to the *New York Times* website and read an article by Nate Schweber reporting from Pyramid Lake:

"Since November 2012, dozens of anglers have been reported catching Pilot Peak cutthroats weighing 15 pounds or more. A Reno man caught and released a 24-pounder. David Hamel, 27, of Reno just did the same thing with a pair of 20-pound cutthroats."

I think about traveling to Nevada to report on what's happening, but I already have too much on my plate. Next year marks the 150-year anniversary of the introduction of trout to Tasmania, and there is much that needs reporting. I am also thinking about writing a couple of magazine-length biographies of Bob Behnke and Joe Brooks, since their bodies of work have so recently proved important to me.

Five months pass by, and I receive an e-mail from Bob Behnke's daughter saying her father has passed away. It hits me hard, much harder than I would ever have expected. Rob Sloane, the editor of *FlyLife*, wonders if I might find time to write an obituary; he also mentions that Joe Brooks's great-nephews—brothers Michael and Joe Brooks—are spearheading an effort to produce a documentary about their great-uncle. "Joe Jr. is coming to Tasmania: maybe you two should catch up." Despite my increasing workload, this seems like an opportunity too good to miss.

I take Joe Jr. to the Styx River in Tasmania's Southwest National Park. The stream is set in a primordial wet forest, shaded by swamp gums (*Eucalyptus regnans*), which rank among the tallest trees in the world. And while I'm showing Joe Jr. how to hunt down big wild brown trout in tea-colored water, he shares some unexpected truths about his great-uncle.

"In early life Joe Sr. was troubled and angry, a drunkard and a whoremonger, a violent man who often ended his late-night drinking binges locked in some jail cell for beating some barfly who happened to agitate him. He belonged to a respectable family in Baltimore, Maryland, and was eventually cast out for being too dysfunctional and irresponsible. Fly fishing was his only release—the water spoke to him—so he ran away to a run-down backwoods cabin. He didn't meet Mary until 1947, when he was forty-six. They were both attending an Outdoor Writers of America convention. He had just published his first book—the first ever on bass fishing—and she, in her role as Ontario's Director of Travel and Publicity, was looking for writers to promote fishing in her province. Something clicked between them and the following year they got married. From that time on, Joe really got his life together. His life's story is testimony to the redemptive power of fly fishing and a woman's love."

I also learn that the Dan Bailey who's doing taimen research in Mongolia is no relation to the Dan Bailey of Livingston fly shop fame. A pity; the coincidence would have been poetic.

Joe Jr.'s proposed documentary honoring Joe Sr. makes me wonder how I might be able to honor Bob Behnke. Again I consider the possibility that I should fish Pyramid Lake, but again my workload seems to forbid it.

Greg French

When Joe and I return from the Styx, I open my e-mails and find that the people at Patagonia Books are suggesting I talk to Peter F. Moyer, a lawyer from Jackson, who is railing against the eradication of mackinaw in Yellowstone Lake. Peter subsequently sends me a copy of *The Yellowstone National Park: Historical and Descriptive* by Hiram Martin Chittenden, Captain, Corps of Engineers, United States Army, published in 1895. The book includes an authoritative list of fish planted in the Yellowstone National Park, including "10,000 yearling lake trout [mackinaw] in the Yellowstone River above the falls in 1890." Peter is bewildered and annoyed that this record is rarely acknowledged despite never being discredited.

Peter also forwards me a dissertation by James R. Ruzycki titled *Impact of Lake Trout Introductions on Cutthroat Trout on Selected Western Lakes of the United States of America*, which concludes that "lake trout predation alone cannot extirpate cutthroat trout in the absence of additional ecosystem perturbations."

These two documents prove to be reliable, but no less controversial for that. I guess that many cherished suppositions hang in the balance. But what do people really fear?

I wonder if by digging below the rhetoric that we all use to justify our conservation efforts, we might find that the things that truly matter to us are not as complex or altruistic as we think. By paring back the protective layers I have wrapped around my own ideals, I have been left with the somewhat troubling realization that the overriding reasons for my worldview are not in any way scientific; instead, they are almost entirely spiritual. And now I wonder, *If spirituality really does matter more than rationality, would this help or hinder the conservation movement?*

Dr. Robert Gresswell is a research biologist with the US Geological Survey, the head advocate and public face of the mackinaw-suppression program in Yellowstone Lake. I read that there has recently been a significant jump in cutthroat trout numbers, and that Gresswell is attributing it to the ongoing harvesting of mackinaw. But invasive species tend to boom and crash whether or not humans intervene, and ill-affected natives often adjust and make some sort of comeback. Surely the real question is, why has it not happened earlier?

I e-mail Gresswell, stressing that I am a passionate supporter of the preservation of native fish in native habitats, and he agrees to answer my questions, to help in any way possible.

I ask him if he knows why Chittenden's record is dismissed out of hand. How many adult mackinaw were likely to have been transferred from Lewis Lake to Yellowstone Lake? Whether the examination of three otoliths constitutes a statistically valid sample. Whether strontium and nutrient levels in Yellowstone Lake have changed over time. If anyone has examined the strontium-to-calcium ratios in cutthroat trout living in the lake prior to 1988. If there has been any detailed analysis of the mackinaw's diet. If he can name a single lake in North America where the extinction of cutthroat can be more credibly attributed to the introduction of mackinaw than to hybridization with rainbow trout or the radical modification of habitat.

His reply in mid-December 2014 is this: "These are complex questions and the answers are equally complex. I guess if there were simple answers, you wouldn't be asking me. Bottom line is this will take some time; I will try to finish by the end of the year." We also discuss the possibility of my returning to America to do an interview.

TASMANIA'S WORLD HERITAGE AREA

I was five when I began preparing for my first day at school. Back then, women's magazines came with special lift-out sections of schoolbook labels, typically sets of photos with spaces underneath for students to write their name along with the subject and grade. My brother, older than me, immediately commandeered the series of cars, planes, and trucks. I couldn't have cared less. My sister, also

Greg French

older, asked if she could have the series featuring dogs, cats, and farm animals. No argument there either. (I loved all our pets, but I'd no more want a picture of an anonymous puppy on my exercise book than a picture of an anonymous man or woman.) I wanted all the wild animals: the kangaroos, koalas, elephants, tigers, ostriches, toucans, walruses, and trout.

My hometown didn't share my *innate* love of wildlife. It preferred clear-cutting virgin forests to make wood chips, decimating kelp beds to make alginates, trawling schools of ocean fish to make fish meal. School ended up being a nightmare; most boys loved cars and log trucks, and I was brutally teased for loving dragonflies and frogs. Even the girls thought I was a sissy. Well, most did, but not all.

Toward the end of grade one we got a relief teacher. She gave us a slide show of her Easter holidays, explaining that she had visited a place called Lake Pedder. I marveled at the wide pink beaches, tussocky moors, and towering peaks, and silently wondered if I might ever go there. She explained that she had walked for days to reach this place, through boggy swamps and tangled jungles, carrying her tent, sleeping bag, stove, and food on her back in a battered rucksack, which she had brought along and allowed us to use for dress-ups. I think she mentioned that this place was in Tasmania, but I didn't accept it—it belonged to Africa or South America or, more likely, the Dreamtime. My new teacher had Hemingway appeal, yet, incredibly, she was a woman. Young and very pretty. I was besotted.

Just four years later, Lake Pedder National Park was almost completely inundated by a hydro-electric impoundment. My church had been plundered. Shortly after that, great swathes of the Hartz Mountains National Park were clear-cut. Even parts of the world-renowned Cradle Mountain-Lake St Clair National Park ended up being lost to development.

By the time I was in my late teens, many of Tasmania's remaining wilderness national parks had been given World Heritage status, and this seemed to offer hope that they might finally be preserved in perpetuity, since the state could no longer act alone to remove protections, and even the federal government was bound by international agreements with UNESCO.

Many of my fellow conservationists began arguing that the best way to preserve the World Heritage Area was to keep people out. But by the time I had reached my mid-twenties, I had lost all faith in lines on maps. As a member of various government committees, I had begun to press the case that environmental protections were only as strong as the support they had in the general community, and that the best way of maintaining public engagement was to optimize the public's interaction with wilderness. After all, history showed that there would always be powerful people who were innately opposed to all things wild, and that no amount of compromise would ever satisfy their bloodlust.

A telling example had occurred in the United States. In 1873 the US Secretary of the Interior, Columbus Delano, boasted that he did not in the least regret the "inevitable" extinction of buffalo because the buffalo prevented the Indians from understanding the necessity of agriculture and hard work. Crazy Horse's response was, "You white men can work if you want to.... We do not want your civilization! We would live as our fathers did, and their fathers before them."

Even as a child I was on Crazy Horse's side. Why would you feel the need to work when there was no need? Surely most people would prefer the Indians' way of life?

Perhaps Representative Omar Conger of Michigan best summed up popular sentiment about the buffalo: "They eat the grass. They trample upon the plains on which our settlers desire to herd cattle.... They are as uncivilized as the Indian."

When the buffalo made their last stand in Yellowstone National Park, there were bugger-all left—just twenty-three, if you remember—and even at this sad juncture there were countless published opinions praising the species' annihilation. Almost everyone understood it to be in the natural order of things that superior domestic cattle should replace inferior wild buffalo, completely and utterly. Even naturalists were resigned to the idea that it would be better to have

the last animals mounted in museums than have them harvested by all-too-active poachers for sale to private collectors.

So far as I can see, too much is never enough; at no point during the destruction of wild places or wild beings do people miraculously see the error of their ways and atone. Conservation battles are never won in perpetuity; they will always need advocates—individuals clever enough and compassionate enough to win over those who may not be *innately* adoring of the wild, but are nevertheless not *innately* antagonistic toward it.

In Tasmania, some 80 percent of the population claim to be opposed to the clear-cutting of old-growth forests, yet we have elected an endless succession of pro-logging governments. Most people care about wild places, but they don't care enough, not as much as fly fishers do. I think back to the conversation I had with Ted and Hugh in Dan Bailey's fly shop. Yes, the reason fly fishers care more deeply than most about wild places is that wild places are a vital part of our daily lives; even when we are not on the water, we are thinking about being on the water. In the battle of hearts and minds, hearts like ours always have the upper hand.

Over the years I have come to passionately believe that if we fail to properly foster and maintain passionate and complex interrelationships with nature, our hard-won protections will end up counting for naught. And as if to confirm my lament, the radio is right now reporting that a newly elected state government has reaffirmed its vow to rescind large parts of our World Heritage Area, including the magnificent upper Styx Valley that I recently fished with Joe Brooks Jr. Not only that, but a newly elected federal government has reaffirmed its vow to assist the state government in the destruction of our natural assets, of our trout habitat.

It seems that our politicians no longer care about international obligations. They don't even care that old-growth logging is costing the taxpayer an arm and a leg. For them, there is something more important at stake: wilderness cannot be allowed to persist; it undermines society's grasp on the necessity of industry and hard work.

For them, when there are only twenty-three swamp gums left, that will be twenty-three too many.

I can't abide the voices of our politicians a moment longer, so I switch the radio to a music station. Someone is interviewing Megan Washington, a gifted Australian songwriter with a singing voice as smooth as a sheltered pool, as thrilling as a rapid. She discloses that she has struggled with a stutter I never knew she possessed. "The only way I can freely express myself is by singing," she says. Melody, she points out, eclipses one's insecurities, bypasses rational doubts, taps the spirit, fills you with confidence and pagan freedom.

Fishing is my way of singing. It is spiritual and liberating. Suddenly I am seized by a compulsion to revisit my remaining childhood temples, places like Lake Sorell and the Salmon Ponds.

LAKE SORELL

I was six. My father and I had just spent what seemed like an eternity driving through barren pastures and dry woodland and suddenly we were at Lake Sorell and there were countless cars, boats, caravans, tents, generators, and a host of very happy and excited campers, mostly anglers with their families and friends.

Straight away we launched our dinghy and began trolling around the open water in the middle of the lake. The fishing was slow, so I spent my time marveling at the large, dense schools of baitfish.

The baitfish—cigar-sized golden galaxias (*Galaxias auratus*)—were endemic to the upper River Clyde system and especially common in its two large headwater lakes (Crescent and Sorell). For more than a century they had been the primary food of the local trout, but they continued to thrive.

I hooked my first trout, a small one, about midday, but I didn't realize until I checked to see if my lure had become fouled with weeds, by which time my catch was already well and truly dead.

Greg French

Later I hooked two much more lively fish. I didn't understand it at the time, but Sorell was already becoming a vital part of my life.

The Silver Plains campsite, happy though it was, proved to be too busy for me and Dad. We ended up finding a more secluded site under a huge cabbage gum near the old weatherboard accommodation house at Interlaken, which had originally been built by the government in the early 1900s but was now unaccountably abandoned. To me it hinted of decay and loss, like a hotel in a ghost town in a western movie. I was scared of it.

For many years, despite its naturally milky water, Lake Sorell was the only lake I fished. The gnarly eucalypts and harsh rock outcrops flanking the shores made me think it was immortal. Why not? It had existed for thousands of years, long enough for an isolated population of baitfish to evolve into something unique.

Despite the lake's oceanic feel, there was less water in Sorell than I imagined. Even when full to brimming, the average depth was only about nine or ten feet.

I was twenty-two when my father was diagnosed with cancer, and suddenly it was my turn to take him to Lake Sorell. Late in the afternoon on the last day of that long weekend, he caught a big brown trout, silver like a sea-runner. He tenderly ran his fingers over the fish's skin—as smooth as the bark on the gum tree by which we camped—and kissed his catch.

We both knew that it was the last trout he'd ever land.

I fished Sorell hard in the years after my father's death, and learned to use flies. I mastered the use of wet flies for catching frog feeders in marshes too shallow and weedy for trolling. I learned that all those rising fish, which refused to eat metal lures, were suckers for dry flies.

During this time I became an adult—and learned to ignore warning signs. Even as heat and drought killed off other smaller bodies of water, I refused to believe that anything would happen to Lake Sorell. How could it? It was one of Tasmania's most popular trout fisheries, accounting for 20 percent of all the angling effort in the state. It was unthinkable that it could die.

I began to notice a decline in the number and condition of brown trout in the early 1990s. The problem proved to be the result of a succession of years when winter flows in the main tributary were insufficient to permit the effective spawning of eggs and survival of fry—a phenomenon that resulted in the lake carrying a diminishing population of old, lean fish. Rainfall soon began to fail at other times of the year too—first in spring, affecting the recruitment of rainbow trout, then in summer and autumn, affecting water levels.

In 1995, European carp (*Cyprinus carpio*) were discovered in the system, which was alarming because Tasmania was at that time assumed to be carp free. The carp management regime initially revolved around attempts to keep the water below the weed line when temperatures rose early in summer, mainly because carp prefer to spawn in marshes and tend to do so when water temperatures warm up.

Unfortunately, the ecological problems associated with maintaining artificially low levels were aggravated by ongoing and unprecedented drought. Severe water-level problems became apparent in 1997, and what remained of the angling community left overnight.

The drought created relentless pressure to draw off more water than ever before for downstream agricultural and domestic requirements. This was easily agreed to; the maintenance of low levels was an official objective and there were no anglers left to argue the toss. Then the foresters moved in and turned the old-growth forests into plantations, which sucked up most of what little rain continued to fall. It is unthinkable that this could have happened when there were more than 8,000 passionate anglers to advocate against destroying the delicate highland catchment.

In 2004, on the twentieth anniversary of my father's death, I drove back to Lake Sorell. On the radio, a "climate skeptic" opined that environmentalists weren't really concerned about the environment or people, they were just waging an ideological vendetta against capitalism and consumerism. "How come a lame-arsed idea invented in the early 1990s has so quickly become an accepted truth?" he frothed. "Where's the proof?" The man's ignorance shocked me. For crying out loud, my primary school teacher talked to us about climate change on the day Neil Armstrong set foot on the moon.

When I pulled up at Silver Plains, my anger dissolved into almost unbearable sadness. The plains were silver, all right. Not with light reflected from wet marshes, but with dusty silt as far as I could see. It was the height of the fly-fishing season yet there was no one there, no one at all. The silence was absolute, ghostly.

How could it have come to this?

Nobody associated with the carp eradication program seemed to understand that on mainland Australia, and around the world, carp do best in water that has been allowed to become intolerable to other fish species. They didn't understand that if, in order to harvest carp, you had to seriously modify the catchment, you were bound to do more harm than good. Ideology, and perhaps fears about job security, seemed to trump pragmatism.

I suppose if you don't love wild places as much as fly fishers do, there is nothing immediate at stake when the bush begins to die. And if you live in the city and only go camping now and then, it is hard to differentiate cyclic changes from devastatingly permanent ones. But those of us who live in the bush day to day, season to season, year to year, decade to decade, become acutely aware of abnormalities.

I noticed that the ancient eucalypts along the lakeshore were sick with dieback. The high branches were dead, and clusters of juvenile leaves were sprouting desperately from lower down on the trunks. The silence, I realized, was caused by more than the absence of people and frogs: it was the sound of the few remaining native trees panting with thirst, gasping their last breaths.

I walked miles around the lakeshore to the big cabbage gum near the accommodation house. My dear old tree was dead, the trunk gray, dry, cracked, and weathered. Actually, on closer inspection, it wasn't quite dead. On one side a thin cord of shiny living bark went halfway up the trunk and ended in a meager flush of bright leaves.

I ran my fingers over the strip of new bark, smooth as trout skin, and kissed it.

SALMON PONDS

On summer evenings, the Salmon Ponds radiates English tranquility and is easily loved. On cold autumn mornings, it is more reminiscent of the fens and moors. But then, it was based upon plans of the famous Stormontfield Ponds in Scotland.

I walk on manicured lawns and blow warm breath into cold, cupped hands. The atmosphere is damp, thick, and gray. I can see the trunks of exotic trees, many planted more than a century ago, scattered like sentries over the grounds, but I can barely make out the hatchery building less than a hundred yards distant. Even the splash and heave of fish and water is muffled by the dense cold and mist.

I walk alongside the grassy banks of Long Pond—a hundred yards long, to be exact—past the interpretation signs that proclaim the success of the original shipment of salmon and trout from England in 1864.

In the early days, the fish in the earthen ponds were brood stock, as irreplaceable as they were valuable. Nowadays, the fish are primarily for public display. Each pool is home to a specific species, not just Atlantic salmon and brown trout but also American fish like rainbow trout and brook char. And oddities like hybrids and albinos.

Native versus nonnative, wild versus domestic, natural versus contrived. In essence, I suppose, the love Peter Moyer has for mackinaw in Yellowstone Lake is no different from the love I have for brown trout in Tasmania.

I first came to the Salmon Ponds as a small child, as most Tasmanians did. Much later, in the mid-1980s, the Commissioner

Greg French

of Inland Fisheries, Rob Sloane, offered me a job as an interim hatchery manager. The opportunity to work so close with live trout seemed like a dream come true.

What I loved most was collecting eggs and milt from wild trout spawners in lake tributaries on the Central Plateau—that and the restoration of the historic buildings and grounds—but I also experimented with hybrids. It seemed like a fun thing to do, and anyway I was curious. Crossing brown trout with brook trout was harmless enough (the offspring are infertile), but I had no idea of the risks I was taking when I began crossing brown trout with Atlantic salmon.

Part of my work also involved the creation of sterile triploid trout, which in some environments are capable of growing faster and larger than ordinary fish. Most of our triploids were on-sold to commercial fish farmers, but some were earmarked for release into small lakes for recreational fishing. Initially the thought of catching giant trout—monster trout—from some of my favorite waters inspired me, but where would my efforts lead?

For a variety of reasons, sterile females are better than sterile males. So we began feeding normal rainbow fry with feed impregnated with testosterone. Over a few years the females developed functional testes. When they were ripe I disemboweled them, ran their gonads through a kitchen blender, and collected cupfuls of all-female sperm. By fertilizing eggs with such a brew, and then heat-shocking them, I was eventually able to manufacture all-female triploids.

Living eggs are translucent, and as the embryos develop you can see black eyes wriggling within. Eyes. Eyes. Eyes. All-seeing eyes. Black as charcoal. Amid a sea of nuclear orange. Literally, my work became the stuff of nightmares.

Perhaps, in my heart of hearts, I always knew that I was creating milking cows and lambs, the insipid strangers I never wanted on the covers of my schoolbooks. So I ended up leaving the Salmon Ponds in order to return to the wild, to work with Rob Sloane on a trout fishery management plan for the Western Lakes Wilderness World Heritage Area.

Publically owned hatchery operations in Tasmania are even worse now, with serious production of trout undertaken in a fully self-contained recirculating factory. Here, modern-day hatchery workers rarely even see the fish; they simply flick switches, twiddle knobs, and read computer screens.

Yet, as Behnke said, although hatcheries may have caused most of the problems that now plague recreational trout fisheries world-wide, their enlightened use is sometimes essential in bringing back some species, subspecies, and races from the brink of extinction.

Greg French

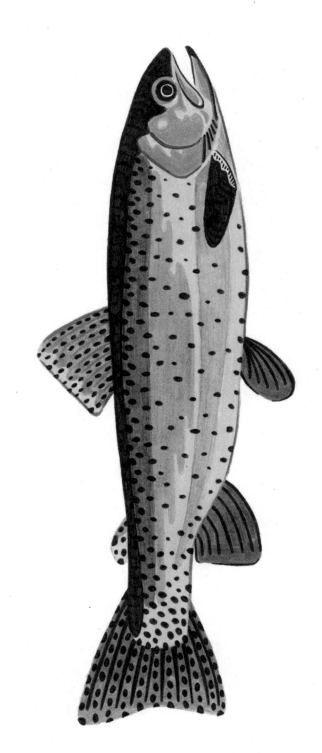

Lahontan Cutthroat Trout
(*Oncorhynchus clarki henshawi*)

Chapter 13: Lahontan Cutthroat Trout

PYRAMID LAKE

Nevada's Pyramid Lake has called to me again, even though I never thought it could. I had hoped to give myself an excuse by going first to Montana to interview Gresswell, but he never did get back to me. I came anyway. Not just because the lake had recently become one of the world's great conservation stories, or even that the story starred Bob Behnke. I think my return also had something to do with my desire to properly understand the irony of hatcheries. In any case, I never for one moment expected that a dedicated sight fisherman like myself could revel in the Pyramid experience, but it's been three days now and I'm loving it.

I'm in a desert on an Indian reservation—standing on the third step of a four-step ladder, water lapping at my knees, fifty yards from the shore—and I've hooked up yet again, this time into a *very* large trout. I know it's a Lahontan cutthroat, but I'm hoping it has Pilot Peak ancestry rather than Summit Lake ancestry.

Frances, atop another ladder five yards to my right, groans melodramatically. "I haven't had a touch all morning."

All morning? First light was only forty minutes ago.

Bill Ladner—a retiree and one of the more successful regulars—stands on another ladder to my left. "Greg, stop hogging all the fish."

Kermit, to Bill's left, turns toward us and is about to add a witticism of his own, but stops mid-sentence when a trout nearly pulls the rod from his hands.

"You're supposed to be watching your indicator, not fishing by Braille," Bill laughs. "Are you guys going to let *any* fish get through to me?"

"No," says Frances triumphantly. I turn toward her, and although momentarily blinded by the just-rising sun, I can tell that she too has "finally" hooked-up.

Rob Anderson, our guide, is wading toward me with a big silicone-meshed landing net.

"You should have called your business 'iNetMore,'" says Kermit.

It's common knowledge around these parts that Rob's clients always catch more fish than everyone else, and that he's gone three years without being skunked. Rob credits his success to his secret flies, but he takes special care with leader length and is fanatical about mending line against the sometimes-strong lake currents.

We are at The Nets—the expansive beaches on either side of the spawning channel, one-third of the way up the western shore. Rob recommended fishing here because big schools of pre-spawners had begun agitating along the flats. "With so many anglers catching so many fish, you won't get a better chance of photographing a Pilot Peak trophy."

I voiced concern about standing on a ladder all day, but Rob said it was essential. "It's like the 'picket fences' at the mouths of lake tributaries in Taupo, New Zealand—once in line, you have to stay put. We've only just come out of winter, and the water will remain—I don't know in Celsius—maybe two to five degrees throughout March into April. When the waves get up, standing chest-deep isn't just cold, it's all but impossible. By standing on a ladder with your torso out of the water you can stay warm all day long, providing you wear waders and multiple layers of warm clothes."

Rob went on to explain the evolution of the ladder line. "In the 1960s, a few people experimented with wooden ladders, but they were too buoyant for stability, and bobbed onto their sides whenever you hopped off to land a fish. Then someone came up with this idea of standing on a metal milk crate. He even attached a string and duck decoy so he could find it whenever he hopped off and pull it to the surface when it was time to go home. Then, in the late 1970s, cheap aluminum ladders became common in hardware stores, and the rest is history."

I asked Rob to show me all the local fly-fishing methods. The most active style involved using a team of two wet flies in conjunction with a weighted shooting head, an eight-weight rod, and a stripping basket. The most productive style involved the use of a bobber-style indicator with a team of two weighted nymphs and a six-weight rod. "The critical thing is that the point fly is small and slender and hovers a couple of inches above the sand," he said.

On day one, despite a clear sky and calm water, Frances and I caught more than a dozen two- to five-pound trout. On day two Frances couldn't join me, so I got Rob and his buddy Chris Evison to fish in her stead. Under a cloudy sky, casting into a perfect northerly breeze, I landed another dozen fish, while Rob and Chris landed at least forty. Some of our catch were Pilot Peak stock, but none quite broke the ten-pound mark.

The plan today is to make sure Frances catches some quick fish at The Nets before we move on to other areas. Apparently fishing from the banks at The Rocks—an area noted for its dramatic tufa formations—can be brilliant. "Chris landed a seventeen-pound fish just two days ago."

The sun is higher now and everything in this vast salt lick telescopes toward me. The desiccation of the Lahontan basin continues to this day. Although it's cold now, daily temperatures in summer commonly exceed ninety or even a hundred degrees Fahrenheit, and it almost never rains. The only water arrives on the southern shore via the Truckee River—whose source is the high Sierras to our west where the winter snowpack can be ferocious—and there is no outflow.

In very snowy years, the lake may rise a few feet, but during the last four years of drought the lake has dropped fifteen feet, continuing the general trend. Tests reveal that water quality has remained stable of late, but stable doesn't mean benign. It is certainly more saline and alkaline than it was before 1844 when the first paleface laid eyes upon it. Yet it is remarkably clear, albeit with a turquoise tinge. I have no trouble seeing the firm, gray silt bed three or four feet below the surface.

I can also see fish swimming left to right a few yards in front of my ladder. Some are moving in small pods, others in great schools. Farther out are a few "rollers"—fish of eight to fifteen pounds that breach once, then disappear. Also beyond casting range are a few very subtle risers and slicing fin tips.

Frances is now playing yet another trout, and Bill is perplexed at her success. "Must be something to do with the way you're holding your mouth."

"The only thing I'm doing different with my mouth is keeping it closed sometimes."

Everyone within hearing distance laughs, especially Bill.

I should be concentrating on my bobber, but I'm distracted by the pastel beauty of the desert. American white pelicans fly overhead, and a loon and two coots swim close by. The hills are turning from purple to pink to gold. On the far shore I can make out the pyramid-shaped island that lends the lake its name. On the hill behind it is a white tide line—some eighty feet above us—marking the level the lake maintained prior to the construction of dams and diversions on the Truckee in the early 1900s. Another tide line 450 feet above that, ancient and faint, marks the high-water level of Lake Lahontan.

The sun rises higher, the breezes fail, the fishing lulls. By midmorning, Rob is wading back to shore to prepare an early lunch. Half an hour later I hear him yelling, "Come eat!" He's calling to Frances and me, but half the locals fall in behind.

Rob remains busy around the gas barbecue perched on the tailgate of his pickup. "Today we have homemade chicken soup and beef burritos."

Joe, a retired guy with blue shades, blue tats, and blue iPad, is keen to show me photos and says, "I just got back from Jurassic

Lake in Argentina." Other people are talking about the best way to customize ladders, and someone else is stressing to Frances that it was Rob himself who popularized bobber fishing in the 1990s. Kermit asks Bill how his wife is, and things aren't so good. Neither are his kidney stones. Frances observes that the ladder line is really a kind of men's shed, and then has to explain what she means.

"You got photos of a Pilot Peak trophy yet?" Joe asks.

"No," I admit.

"Well, the day's young so you're still in with a good chance. Since early December, Bill's landed thirty-five fish in excess of ten pounds, including a couple over twenty pounds. Truth is, all of us here have taken fish from twenty to twenty-five pounds, and it's only a matter of time before Pyramid Lake gives up a thirty-pounder. They're precious, these Pilot Peak fish."

After lunch no one in the ladder line is having success, and Rob suggests it might be a good time to visit some other fishing sites around the lake. We end up going north to The Rocks, then south to a museum near the inflow of the Truckee. And on the way back we stop in to talk with the Paiute who run the hatchery.

Do you believe you are what you eat? Many American Indians seemed to think so. The Shoshone people of the Snake River below Shoshone Falls were known as the Salmon-eaters, a band of several hundred Western Shoshone were the Sheepeaters, the Northern Paiute of Stillwater Mash were the Cattail-eaters (Bulrush-eaters). I assumed that the Paiute of Pyramid Lake would be the Trout-eaters, but apparently that title belongs to the Winnebago people of Wisconsin. The Paiute of Pyramid Lake are the Cui-ui-eaters (pronounced **kwee**.wee).

Cui-ui are a sucker fish endemic to Pyramid Lake, usually blue-gray or black in color, which can grow to six pounds and were historically more abundant than cutthroats. Nonetheless, cutthroat trout were always revered as food, and today they are revered as a tangible link to the traditional and wild past.

The Paiute place great importance on the maintenance of wild fish stocks. Natural selection beats artificial selection every time, they believe, and by way of example I am reminded that the domestic Lahontan cutthroats stocked annually into neighboring Walker Lake were ultimately unable to adapt to rising alkalinity levels. Where that leaves the Pilot Peak stock in the long term is anyone's guess, especially if Pyramid Lake continues to desiccate.

A dream of many locals is that the Pyramid fishery will eventually be maintained entirely by natural recruitment from restored spawning grounds in the lower Truckee. But even though the Paiute have been successful in having more water returned down the river, it is unlikely that the lower reaches will ever again be cool and fast enough to provide year-round trout habitat. In any case, the healthy parts of the Truckee (through Reno all the way to Lake Tahoe in the Sierras) teem with nonnative trout, including rainbows, which, as we know, commonly hybridize cutthroats to extinction. It seems to me that the only way that "safe" natural spawning might ever be made possible would be by constructing an elaborate artificial spawning canal using either water piped (and screened) from the healthy parts of the Truckee or, in a more contrived system, pumped from below the thermocline of Pyramid Lake.

After leaving the hatchery, we drive back to The Nets, and Rob notices Bill and Kermit in the shallows manhandling a huge landing net. "They must have just caught a big one!" he exclaims, and guns it. I jump out of the pickup before it has fully stopped, and finally manage to photograph a truly grand Pilot Peak cutthroat. "Thirteen pounds," Rob declares, reading from the scale on the net. And we launch into a melee of high-fives and man hugs.

This is Behnke's legacy. During his lifetime, the world's leading authority on salmonids attracted plenty of skeptics, but his uncanny intuition trumped conservative outlooks every time. And as important as his insights have been for Pyramid Lake, they are even more relevant in Yellowstone, Mongolia, and Tasmania. In fact, they are relevant everywhere wild trout are found.

RENO AND THE TRUCKEE

A bit before the end of the day we ask Rob to take us back to our hotel. "You're not enjoying yourself?" he asks, dumbfounded.

"Come on, you know we're having a great time. It's just that this is our last night in Reno and we want to have a go at catching some of the dozens of big pre-spawned rainbows we've seen milling about in the currents below the main road bridges in the city center."

"You can see those fish?" He seems surprised. "I guess you might, with the river being so low and all."

It's hard for out-of-towners like Frances and me to imagine that the winter has been as mild as everyone is saying. Flying in from Los Angeles over the Sierras, the snowpack seemed vast and deep. I concede, though, that as we approached Reno it became impossible to tell for sure what was snow and what was salt.

From the air, Reno seemed to cover way too much area for its population of about 200,000. The basin in which it was set was flat and expansive, and by far the most dominant feature was the towering white Grand Sierra Resort. Our hotel.

After we say our goodbyes and thank-you's to Rob, we wander through the lobby to the lift. As in all casinos hereabouts, the lobby doubles as the main gaming area. There are countless dozens of slot machines, all themed—cartoon characters, pop singers, board games, anything prominent in popular culture. The players strike me as robotic and sad. Yet Rob and many of the people I liked from the ladder line had talked to us about how fond they were of "casino culture," of going to a show and having a little flutter on the craps tables. "It's a cheap, fun night out."

After a quick freshen-up, Frances and I walk back out onto the streets. I carry a six-piece rod in a small aluminum tube and a few flies in my daypack. I haven't bothered with waders, so the mile-long stroll into the heart of the city should be pleasant.

There are billboards everywhere along the highways, but I'm comforted by the fact that many display socially progressive messages, asking people to donate to the homeless or contribute to arts projects.

The streets themselves are many lanes wide and flanked by concrete sidewalks, concrete walls, and concrete buildings. Casinos cover whole blocks, except for a number of little alcoves that shelter quaint wedding chapels. Reno's not as garish as Vegas—seventies music isn't being piped from plastic flowerbeds and there are way fewer Elvis impersonators and religious zealots—but there's not much in the way of living trees or grass. Nor is there any central shopping district, restaurant strip, or community hub. Instead, retail outlets are mostly big barns set miles apart. Perhaps I shouldn't be surprised that we are alone on the sidewalks.

Where will we eat after we finish fishing and photographing? The casinos offer genuinely good food, including excellent sashimi and sushi—I wonder if the Japanese food culture can be traced back to a desire to cater to Asian high rollers—but we don't really want to have dinner in yet another gambling joint. Unfortunately, there is little else on offer, not unless you count franchised fast-food outlets. On previous evenings we have found the odd little oasis, but they are few and far between and tend to serve weak "espresso" by the pint and in paper or plastic cups.

We look for a place to fish beneath the bridge on Second Street, but a large group of homeless people have set up a semi-permanent camp and we don't want to intrude on their personal space. Overt poverty like this is shocking to our Australian eyes, or at least the magnitude of such poverty. We end up moving to the bridge on Sierra Street. The river here is bound by concrete walls, but there are narrow esplanades of exposed shingle from which we can easily cast to the clusters of trout we spot from the bridge. A small weighted nymph in conjunction with a small indicator does the job nicely. The trout are large, in prime condition, and super strong, and although I

Greg French

vowed to catch only one or two before dinner, I'm finding it difficult to leave. Nonnative or not, these fish are to be cherished.

Dusk is fast approaching and bright neon lights are already reflecting rainbow colors from the riffles and glides. Inside the glitzy buildings flanking the river, thousands of people are gambling. Here on the riverbank, tens of yards away from those buildings, Frances and I are utterly alone. For most people, it seems, contrived environments are superior to undisciplined ones.

Chapter 14: Endpieces

ANTHROPOCENE

There is no doubt that human activity has dramatically affected Earth's atmosphere, soils, and biodiversity. Many scientists are now using the term "Anthropocene" to identify the time when the effect of human activity became pervasive. The problem is not whether the idea has merit, but of deciding when the Anthropocene began. In the book *Plows, Plagues, and Petroleum: How Humans Took Control of Climate*, noted palaeoclimatologist William F. Ruddiman argues persuasively that the date is 7,000 years ago. At this time, he says, the atmospheric change resulting from deforestation for agriculture reached a tipping point that stopped planet Earth from accelerating toward another glacial maximum.

If Ruddiman is correct, humans could be directly responsible for the retreat of glacial ice from the Yellowstone Plateau 6,000 years ago that exposed Two Oceans Pass and enabled Yellowstone cutthroat trout to invade the Yellowstone River. In other words, it is possible that the trout in the Yellowstone catchment exist "only as the result of human agency," the very same argument that is used to argue the illegitimacy of mackinaw.

When I was a child, I was taught that Australia's only wild dog, our distinctive dingo, had been brought from Thailand to Australia 6,000 years ago, and at the time many respected naturalists argued that it was nonnative vermin. These days, the problem of whether to protect it is compounded by the fact that most dingoes on mainland Australia have hybridized with domestic dogs.

Examples of human-induced change are everywhere. It is reported that pythons in northern Australia now have much smaller heads than they had prior to the introduction of cane toads, presumably because smaller heads make it harder to ingest lethal doses of toad poison. Numerous bird species that live near roads have smaller wingspans than they had prior to the invention of the automobile, presumably because small wingspans offer greater maneuverability and lessen the chances of becoming roadkill.

None of this is to say that humans are the major cause of evolutionary change, any more than one could say that any other species is the major cause of evolutionary change. Our own evolution, for example, has been decisively influenced by the dog and beasts of burden like the horse and bullock. Without animals, civilization as we know it may well have been impossible.

We are not separate from nature—we are a part of it.

The last lava flow in Yellowstone National Park occurred 70,000 years ago. Yellowstone cutthroat trout became isolated above Shoshone Falls in the upper Snake River 60,000 years ago. They invaded the Yellowstone catchment just 6,000 years ago. The next major volcanic eruption, long overdue, will almost certainly exterminate the Yellowstone cutthroat trout from all of its natural range. It will be an entirely natural process. So, should we reintroduce the species after such an event? If we had the means, should we prevent such an event from happening in the first place? How would these actions fit with the aims of preserving natural diversity? Of allowing nature to run its course? Of fostering respect for the well-being of animals?

What I am suggesting is that, just as there is no way to define *species*, there is no way to define *natural* or *unnatural*, *native* or *nonnative*; and worse, the aims of preserving biodiversity are often at odds with our ideals for animal welfare. We have a quandary: conservation is more complex than most of us are able to comprehend, certainly more complex than most of us *want* to comprehend.

Despite all this, I desperately care for the preservation of "native fish" in "native environments." The question is, why? Here's the best I can come up with:

These days, when we are hurtling at an accelerated pace toward the contrived and bland, diversity and serendipity come at a premium. Things primal are essential to my sense of self; I need wild places—where the rate of change is tolerable—as a spiritual harbor.

Some people say fishing is justifiable only if absolutely required for food. This is akin to saying that sex is justifiable only if absolutely required for procreation. All I am able to offer in my defense is this: *The Great Spirit has put in their hearts certain wishes and plans, and in my heart he put other and different desires. It is not necessary for eagles to be crows.*

Those who argue that discouraging fishing would make people more "civilized" and help secure the well-being of the Yellowstone trout ignore the fact that without anglers, there would be few people who even knew that fish populations were in trouble, let alone advocate for their protection.

I think of the girl in Gardiner, how she understood that transit passengers entering America for a few minutes or hours often left with neutral or negative feelings about her homeland, whereas tourists who spent weeks interacting with American people and landscapes ultimately learned to love her country.

Of all the places I've visited in the world, I think Yellowstone is the one most like a touchstone. Surely this is the place that other places should be compared to in order to test the veracity of all we deem precious.

THE MADISON RIVER

In July 2015, I spent a couple of weeks hiking in California's Golden Trout Wilderness, and while in America it made sense to revisit Yellowstone to see how things were faring. I had hoped to take this trip with Frances, but she confessed that after our Cache Creek experience she really didn't feel comfortable camping out in bear country anymore. Longtime friend Ric Dowling ended up coming in her stead.

First stop was the Madison where, in order to assess the mountain whitefish situation, I organized a drift down the Fifty Mile Riffle, choosing Blue Ribbon Flies from among other reputable West Yellowstone outfitters for no other reason than it was owned by Craig Mathews, author of *The Yellowstone Fly-Fishing Guide* (with Clayton Molinero) and *Fly Fishing the Madison* (with Gary LaFontaine).

Unfortunately, Craig wasn't working when we arrived, but his staff were excellent. No sooner had I entered the shop than a young a man came around from behind the counter and offered his hand. "You must be Greg and Ric. I'm Peter Scorzetti. I see from the booking sheet that you want to fish the Fifty Mile Riffle for whitefish. Unusual request. I wish I was going with you. But I'm booked out, so you'll be going with Ivan."

Ivan had to drive me and Ric well downstream below the Hebgen Dam. "The uppermost part of the Fifty Mile Riffle is reserved for wade fishing. We can't launch the drift boat until we get to Lyons Bridge. From there we'll drift for maybe fifteen miles before pulling out at Ruby Creek."

"Will we be able to polaroid the whitefish?" I asked.

"Not a chance. I'll set you both up with two-nymph rigs under a bobber float, and we'll prospect the currents. Are you sure you don't want to dry-fly fish for trout?"

"What are our chances of actually catching whitefish?" I persisted.

"Should be easy. They are everywhere in the currents. They don't often rise to dries, but they are suckers for deep-fished nymphs. The trout are completely different. They hold in specific types of habitat, and although high temperatures in July mean that we might not see many risers, we can easily tease them to the surface. It would be much more fun than nymphing."

Ivan didn't seem to know much about the reported disappearance of whitefish in the early years of the new millennium. I couldn't work out if his lack of knowledge stemmed from his disinterest in whitefish, disinterest in nymphing, or if the fishery had recovered before he started guiding. My understanding was that during the worst of the whirling disease epidemic—when there were bugger-all rainbows—the angling presence had collapsed, so I guessed the new cohort of adherents probably didn't enjoy the advantage of an un-broken, generations-long knowledge of the fishery's ups and downs.

Eyeing the twenty-two drift boat trailers in the busy Lyons car park, I asked if this was normal. "An average of forty boats put in each day, but there can be sixty or more in the peak months of June, September, and October," Ivan said.

Within an hour of setting off, we had hooked and released half a dozen solid whitefish, all weighing two pounds or so, all good fighters. And then, to Ivan's apparent relief, Ric and I finally agreed to fish for trout.

"To be successful, you have to concentrate on maintaining a natural drift over gravel bars in distinctive seams behind emergent and semi-emergent boulders," he said.

Most trout were sprats of eight to thirteen inches that performed splashy rises to our tiny white-post emerger-style flies and were sometimes difficult to hook. But occasionally there were more seri-ous gulping takes from eighteen-inchers (fish weighing more than two pounds). We even managed to boat a couple of twenty-inch rainbows, which caused Ivan to become quite animated. "Rainbows max out at twenty inches in the Riffle, and getting two in a session is really good going. Browns can reach twenty-two inches, almost four pounds, but they are even rarer. My clients might land only one a week," he told us.

The size of the rainbows didn't excite me as much as their rela-tive abundance. Even better, the older fish had obviously spawned more than once, so it seemed likely that the species' resistance to whirling disease had become entrenched. Woo-hoo! Maybe a similar resilience had brought the whitefish back from the brink as well, and perhaps, after all, the Madison fishery would continue to provide champagne fishing for decades to come.

SODA BUTTE CREEK

By 2015, everyone seemed to agree that it was the cleanup of the McLaren Mine tailings dam near Cooke City that enabled the nonnative brook trout to invade the Soda Butte. Apparently the dam (on a small headwater tributary) had previously acted as a chemical barrier against downstream migration into the uppermost part of the Soda Butte's main stem, demonstrating yet again that we should always remain mindful of unintended and unpredictable outcomes when initiating conservation works.

Also by 2015, researchers working on the brook trout eradication program had concluded that electrofishing had become "ineffective"; the number of brookies killed during each field trip had flatlined, suggesting that repetitive electrofishing could not reduce the population below a certain low level. (Who would have guessed.) Consequently, they decided that rotenone should be used to poison the entire twenty-eight-mile length of the Soda Butte upstream of the Ice Box Canyon. This was bound to involve the killing of thousands of native cutthroats—two hundred or so for every one brookie—and would likely need to be repeated, but by all accounts it was definitely going ahead.

Ric and I visited the park just weeks before the poisoning was due to begin, and we drove around to the Cache Creek trailhead (on the Soda Butte immediately below the Ice Box) to see how the locals felt about the initiative. I ended up talking to a dozen or more anglers, but strangely enough no one admitted to being aware of the brook trout eradication program. Of much more concern, apparently, was a new law compelling anglers to kill all rainbow trout taken from most of the Lamar system, including those taken from tributaries like the Soda Butte, the upper Slough, and Cache Creek.

One veteran angler was particularly incensed: "Killing wild trout is revolting to me. Me and my buddies have spent decades trying to help foster respect for wild fish, and mandating slaughter undermines everything we've worked for."

I asked him if he thought the National Park Service should have spent more time explaining the problem, or if it should have asked people to take rainbows home to eat rather than kill and discard them. He would have none of it. "We are obliged to put hybrids back if they display a red slash under the jaw, even if they look more rainbow than cutthroat, so the law is just plain stupid."

Since there were so many people fishing the Soda Butte, Ric and I ended up doing a day hike to Cache Creek. We didn't see any grizzlies this time, but we found other reasons to worry, not least of which was the sad state of the conifer forests. Dieback had taken hold in the couple of years since I had last visited, and whole hillsides seemed mostly or completely dead.

On the upside there was a very good rise in Cache Creek. We were in the thick of catching and releasing fish when a trio of hikers crossed the stream a few yards ahead of us and stopped to chat. The youngest member of the group, we discovered, was an off-duty fisheries officer in his late teens or early twenties. He politely introduced his elderly parents, whom he was taking to one of his favorite campsites farther up the Lamar, and then he gently reminded us that we had to kill any rainbows we caught. I couldn't resist asking why.

"Well, it's the law, but more importantly we need to get rid of as many rainbows as we can to stop them hybridizing with the native cutthroats."

"But surely if you really thought that angling could have any meaningful impact on nonnative trout, you wouldn't need to poison the Soda Butte. And since the law is almost certainly going to be completely ineffective, might it not end up doing the conservation movement more harm than good? I mean, you are bound to foster ill will among some wild trout enthusiasts when you force them to act against their convictions."

He laughed and admitted that he hadn't considered this.

Then Ric asked him what he knew about the dieback.

"Two things are causing it: blister rust, which mainly affects whitebark pines, and an explosion of pine beetles. The blister rust is nonnative and we would eradicate it if we could. The beetles are native, so management is more knotty, not that we can do much about it anyway. The thing is, both problems are probably interlinked, and exacerbated by climate change. By the way, where are you going after you leave here?"

When we told him that tomorrow we were going to fish at Buffalo Ford, he gave us some of his home-tied flies and offered some considered fishing tips. Inexperienced or not, I reckon he's just the sort of fellow the National Park Service needs to nurture.

BUFFALO FORD

Rumors were flying that the fishing at Buffalo Ford had rebounded big time, and Ric and I could barely constrain our enthusiasm to get there.

There were a lot more people fishing the river than when Frances and I last fished it, though nowhere near as many as I had feared. Strangely enough, most of the anglers were men in their sixties and seventies, even eighties. And by remarkable coincidence the only young person turned out to be Peter Scorzetti from Blue Ribbon Flies. When I asked him why he thought there were so few novices on the water, he reasoned that most young anglers—a whole generation of them—had missed out on experiencing the upper Yellowstone in their formative youth. "It's a real pity," he said. "History is a big part of the fishing experience, especially if you have actually lived that history. People will rediscover this place, for sure, but for many it will be by accident rather than as a normal rite of passage."

The anglers Peter, Ric, and I observed were clearly very experienced, and most could cast very long lines, but they were struggling to notice subtle rises. Worse, they seemed reluctant to wade too far out into the deep, heavy current where most of the more showy and regular risers were doing their stuff. I guessed that old age had taken its toll on their eyesight, hearing, strength, and agility. In any case Peter, Ric, and I were the only ones who ended up having much success. Over the course of several hours we landed more than twenty cutthroats between us, all twenty to twenty-two inches long and so chunky across the shoulders that the biggest must have weighed well over four pounds. I could scarcely believe how much the fishing had improved.

Several anglers came to my side each time I waded to the bank to beach my catch, and invariably I was asked what fly I was using. I explained that I thought my success was mostly the result of being able to wade out to a position where I could perform short, accurate casts and easily maintain drag-free drift, but I was more than happy to give them some of my generic Australian flies.

"Flies: the currency of anglers," one fellow commented happily as he tied on one of my offerings.

I laughed, and then his nonfishing wife quietly appeared at our side and said, "Don't get to thinking Fred is observant or anything. Why, only yesterday a big grizzly came down to drink just a few yards from where he was fishing, and do you think he noticed?"

To which another man, Jim, added, "You can trust what Martha is telling you. I was a witness to the event."

By this time the rise had slowed considerably, so Fred, Jim, Martha, Peter, Ric, and I retreated to a high bank to see if we could spot fish that weren't feeding at the surface.

We soon found a nice vantage, and almost immediately noticed several large trout shuffling about in the deep currents a long cast from the bank, prompting Fred to say, "It's great how this fishery has been rehabilitated. The cutthroats are back in good numbers, and the average size and condition is better than I've ever experienced. This year you can see good numbers of spawners from the Fishing Bridge too, and I've had some great days at Gull Point on

Yellowstone Lake casting to rising fish, most of which have been feeding on *Callibaetis* and small green drakes."

Jim added, "We sure can be grateful to those scientists who have been culling the lake trout, can't we?"

"I wonder if there's more to the story," I offered before going on to talk about the updated estimates of trout numbers in Yellowstone Lake made by university student John Syslo.

I guess the figures I quoted on the riverbank would have been fairly rough, and perhaps I overegged things by making the years line up, but I certainly didn't exaggerate the trend. (For the record, Syslo's estimates for the cutthroat population were 1.96 million in 1986, 463,000 in 2000, and 1.31 million in 2012, while the lake trout population was zero in 1986, 126,000 in 1998, 746,000 in 2012, and 608,000 in 2013.)

"Any way you look at it," I stressed, "cutthroat numbers seem to have rebounded at the same time that lake trout numbers were dramatically on the increase. Right now there are many more lake trout in Yellowstone Lake than there were when the cutthroat population crashed. Also, the most recent study of lake trout guts found that they were feeding mainly on scud. Not one sample contained a single cutthroat. The fact that lake trout are competing for the same food as cutthroats probably isn't ideal, but even so, I don't think that the evidence supports the idea that lake trout caused the cutthroat crash, at least not by themselves. Syslo claims that the culling program is both effective and essential, but I reckon the pattern of crash and recovery in the cutthroat fishery is more consistent with what would happen if whirling disease was the main problem."

To my surprise Jim pounced on me. "What would you know about this fishery? You live in Australia, don't you? Well, I live in Montana and fish here all the time. There isn't any whirling disease here in the Yellowstone—that's the Madison you're thinking about. You should listen more to the experts. I've got no time for members of the anti-science brigade."

Anti-science? The information I gave was entirely science based. I silently cursed Yellowstone's fishery managers for becoming so preoccupied with a single theory that they had failed to properly disseminate knowledge of all the important things.

Then a fish took up station directly below us and we all stopped arguing and began taking turns at trying to fool it. It was a very big animal, but judging from its scarred skin and deformed jaw it had already been caught several times this summer. Perhaps that's why it kept ignoring most of our dries and nymphs, and inspecting and refusing the rest. But we kept offering each other flies and encouragements, and after countless casts Jim finally hooked up. While he played the fish from the cliff top, the rest of us made our way down to the nearest beach of exposed shingle, and when Peter eventually netted Jim's fish for him we all burst into laughter and enthusiastically slapped each other on the back.

Still, the political and environmental imperatives that underpin our fishing are vital to all our futures, and at some point we really are going to have to take the time to talk about the difficult stuff.

ACKNOWLEDGMENTS

Thanks, as always, to Frances Latham and Rob Sloane—special confidants, fearless critics, most reliable proofreaders.

BIOS

GREG FRENCH is one of Australia's best-known fishing authors. He spends most of his time in Australia and New Zealand, but he has fished extensively in South America, North America, the British Isles, Iceland, Eastern Europe, and Mongolia. He's written numerous books, including a comprehensive guide, *Trout Waters of Tasmania*, originally published in 1984 and last updated in 2011; *Frog Call* (New Holland, 2002) a work of literary nonfiction; and its companion, *Artificial* (New Holland, 2008). *Menagerie of False Truths* was published by Exisle in 2010. French also cowrote with Nick Reygaert the acclaimed documentary *Hatch*. In 2013, Reygaert and French produced a companion DVD, *Predator*, which won the Best DVD award at the 2013 IFTD tackle show in Las Vegas. He lives in Molesworth, Australia.

Illustrator GEOFFREY HOLSTAD is an artist, creative director, plein air graphic designer and citizen meteorologist. He is currently daylighting as an apparel graphic designer at Patagonia in Ventura, California. By moonlight, he is the co-founder and director of Cabin-Time, a roaming creative residency to remote places.

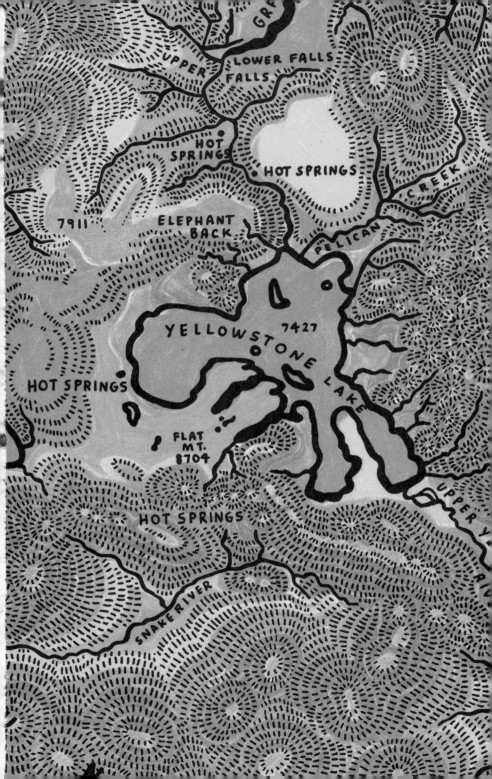

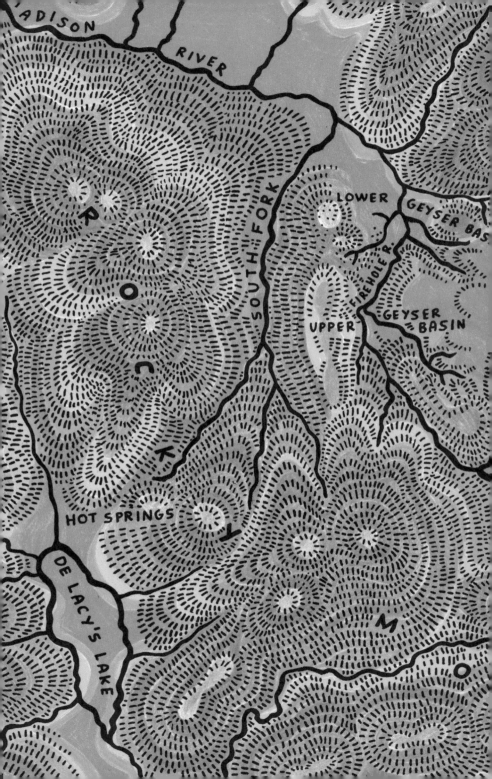